The New Archery

The New Archery

Revised Edition

David F. Butler

South Brunswick
New York: A.S. Barnes and Company
London: Thomas Yoseloff Ltd

Library of Congress Catalogue Card Number: 78-37806

A. S. Barnes and Co., Inc.
Cranbury, New Jersey 08512

Thomas Yoseloff Ltd
108 New Bond Street
London W1Y OQX, England

ISBN 0-498-01142-9

Printed in the United States of America

This book is dedicated to my sons, Donald W. and David E. Butler, who spent many hours measuring angles and distances during the flight-shooting tests.

Acknowledgments

Appreciation is expressed to Mr. E.D. Lowry who has helped the author develop a better knowledge of ballistic theory through many discussions during the past fourteen years. Appreciation is also expressed to Mr. Irving Allen who created *Figure 9* of *Chapter 4* and Mrs. Geraldine A. Patti who put in extra hours not only in assisting with the manuscript but in proofreading. Appreciation is also due to my wife who put up with the inclusion of quantities of archery equipment on our vacation, and many inconveniences during the preparation of this book.

Contents

List of Illustrations

The New Archery

1
Archery Equipment

Archery, as a sport, has had a tremendous resurgence in the past 15 years due primarily to the hunting feats of outstanding bowmen and the development of excellent new equipment. Prior to World War II, the picture was very different. There was a choice between various patterns of wood bows, but the general style and performance of archery equipment had changed very little in hundreds of years. After the invention of firearms, the bow ceased to be a first-line military weapon in the Western countries, and as a result, archery has become more and more oriented toward leisure-time sporting uses.

There was a resurgence of interest in archery in the United States after the Civil War. Unfortunately, the sport developed as a formalized, victorian type of activity to be carried out in formal gardens with lightweight equipment.

Since World War II, archery has achieved a new image, and a much more important place among the outdoor sports. It is a participant sport, and with proper equipment all age groups can have fun, and compete in archery events. As with most other sports requiring equipment, the beginner has a hard job deciding what equipment fits him best. Most people purchasing archery equipment make one serious mistake. They purchase a bow that is much too heavy. Most archers do not participate continuously and the muscles used in archery are not in peak training form. Let's take a simple example. A man asks for a 40-pound bow. This will normally be shipped from dealer's stock anywhere in the 38-pound to 42-pound range. If the archer is of average height, he may easily draw the bow 29 or 30 inches instead of the rated 28 inches. The force to draw a 40-pound bow increases about two pounds for each inch of extra draw. Thus, our archer may find that his proper draw with a "40-pound" bow runs anywhere from 38 to 46 pounds. This is all very easy in the store when the archer flexes the bow several times. He rarely places an arrow on the string in the store, and, therefore, has no way of knowing how far he has drawn. The result is that the bow on trial is usually underdrawn by several inches, and this will reduce the force required by about two pounds for each inch below the 28-inch rated length.

The problem gets worse as the archer really starts to improve his technique. For accurate shooting, the arrow is drawn smoothly to the full draw position. For those with long arms, the most common position is to bring the string in contact with the end of the nose, and the fingers holding the string in contact with the chin. For those with shorter arms, the usual position is to draw the string alongside the cheek, with the fingers in contact with the ear.

For either position, the bow is not drawn and the arrow released immediately, but this force must be maintained for two to three seconds in order to make final sight adjustments before release. This holding is known as the "anchor." An afternoon's practice can easily include 90 shots, which is the number in a standard American round. A beginning archer will find he has an extremely sore set of muscles if a heavy bow has been chosen. As the muscles tire, the natural tendancy is to release the arrows more quickly, before the final sight adjustment has been made. Even if the bow has been properly drawn and anchored, tired muscles will shake and give a poor release.

The principles of archery are similar to those in pistol or skeet shooting. All of these participant sports require good muscular coordination and sufficient practice so that the muscles are trained to function in a smooth pattern shot after shot. Most shooters can detect an easily identified pattern in their shooting. The first scores of the day are relatively ragged because motions are jerky and gross errors result. As you warm up to the sport, and get in stride, then the scores will reach an optimum level for your stage of training. At a certain point, there is a very detectable break in the scores. This is where muscles begin to tire and are no longer able to make the very fine adjustments that lead to really good scores. In archery, the release is less smooth and the final adjustments are less accurate and the scores begin to suffer badly. The occasional archer can rarely get through a 90-shot round using a 40-pound bow without reaching this point of tired muscles and falling scores.

For target shooting up to 50-yards range, a bow weight of 30 to 35 pounds appears to be the best choice for the average weekend archer. This weight is also used by many championship shooters. Even with this bow, the archer will end up with sore muscles the first few times it is used.

If there are young archers in the family, then a bow should be selected to fit their needs. The proper bow should be estimated by an experienced archer to fit the particular children. For teenage girls and women, a bow in the 20- to 27-pound range should be chosen. A bow of over 30 pounds is too heavy for the vast majority of women, and nothing will discourage an archer faster than a bow which is a strain to use. If one bow must be chosen for a family, a 25-pound bow is an excellent choice. It will require only about 22 pounds at a 26-inch draw for older children and women, yet will be up close to 30 pounds in draw if the man of the family has a normal draw of 30 inches.

LENGTH OF DRAW

The beginning archer is faced with a bewildering array of manufacturers and specifications in choosing a bow. One thing that is often overlooked is the limited draw length on inexpensive bows. It is much better to purchase a last years' model laminated bow with a 30-inch draw limit than to select the latest model in a cheaper grade with a draw limit of 28 inches.

The more inexpensive lines of bows generally have shorter limbs, and slightly different characteristics from a long-limbed target bow of good quality. For a woman's bow, a length of 62 inches and a draw limit of 28 inches is usually adequate. For a man, a target bow of 66 inches or longer with a draw limit of at least 30 inches is recommended if the archer is over 5 feet 10 inches tall.

More information on bows is contained in Chapter 2. The advice of an experienced archer can be of great benefit in selecting equipment. Unless a bow is being selected specifically for hunting, then a long-limbed, smooth-working target bow with a draw weight in the low 30- to 40-pound range is a good first bow for men.

HUNTING BOWS

A hunting bow is used for only a few shots in any one

day, and a heavier bow may be tolerated. Many states have laws requiring a bow of at least 40-pounds draw for hunting use. In hunting it is useful to have a compact bow so that moving through brush will be easier, and the game will have less to see as the archer is stalking and drawing. It is extremely difficult to design and manufacture heavy hunting bows that will give a smooth draw over a 30-inch working range. These bows are necessarily quite expensive. The shorter the bow, the more acute the angle of the string with the bow fully drawn. Some hunting bows are as short as 48 inches in length, and the angle of the string when fully drawn is quite severe. This angle tends to crowd the three fingers holding the arrow together, and makes a smooth release more difficult.

A short-limbed hunting bow is a difficult first bow to use in learning good technique. A better plan is to use a long-limbed, lower-weight bow to develop good form, and then select a hunting bow which best fits your shooting form, and type of hunting. Heavy hunting bows are very difficult to string properly. Stringing all modern recurved bows must be done carefully in accordance with the manufacturers recommendations, but a short, heavy hunting bow can pose a major problem in getting the bow strung. Often, experienced archers sit on the hunting bow in a crouch position, but the beginning archer is wise to use a bow-stringing attachment, or secure help from a friend to avoid damaging the bow.

ARM PROTECTION

The arm guard is used to protect the arm, and to provide a smooth surface for the bowstring to strike. The bowstring is travelling at the same velocity as the arrow at the time of release, and it continues forward at high speed, stinging an unprotected arm.

Two arm guards are shown in *Figure 1*. On the left is a commercial model made of tough leather with a steel-reinforced band sewn in the center. These are excellent protection, and last very well. The arm guards are held on with either two or three elastic straps, which can be adjusted to fit the bare arm or over clothing.

If the budget is tight, ordinary shirt cardboards can be folded over double and held on the arm with two heavy rubber bands. An illustration of this arm guard is shown on the right side of *Figure 1*. The cardboard is not very elegant, but it will do the job while money is being saved for better equipment.

FINGER PROTECTION

Some kind of finger protection is a must. For a few shots any kind of leather gloves may be used, but these are really not very satisfactory. Three types of finger protection are shown on the bottom half of *Figure 1*. The most inexpensive type of protection is the shooting tab, or archery tab shown in the right-hand sketch. This is a small piece of stiff leather. The middle finger is placed through the hole in the tab, which positions the tab on the hand. After the arrow is placed on the string, the tab is hooked around the string, with the arrow fitting in the indentation in the tab. Inexpensive tabs consist of a single layer of leather, and they are very difficult for the beginner to use. The tab must be carefully set for each shot, and it tends to slide around on the finger, leading to variations from shot to shot. Better tabs are designed with a double layer of leather, and a separate loop for the finger. They also have a guide in the arrow slot. Good quality tabs are sometimes used by tournament archers, but in general the new shooter is advised to avoid them.

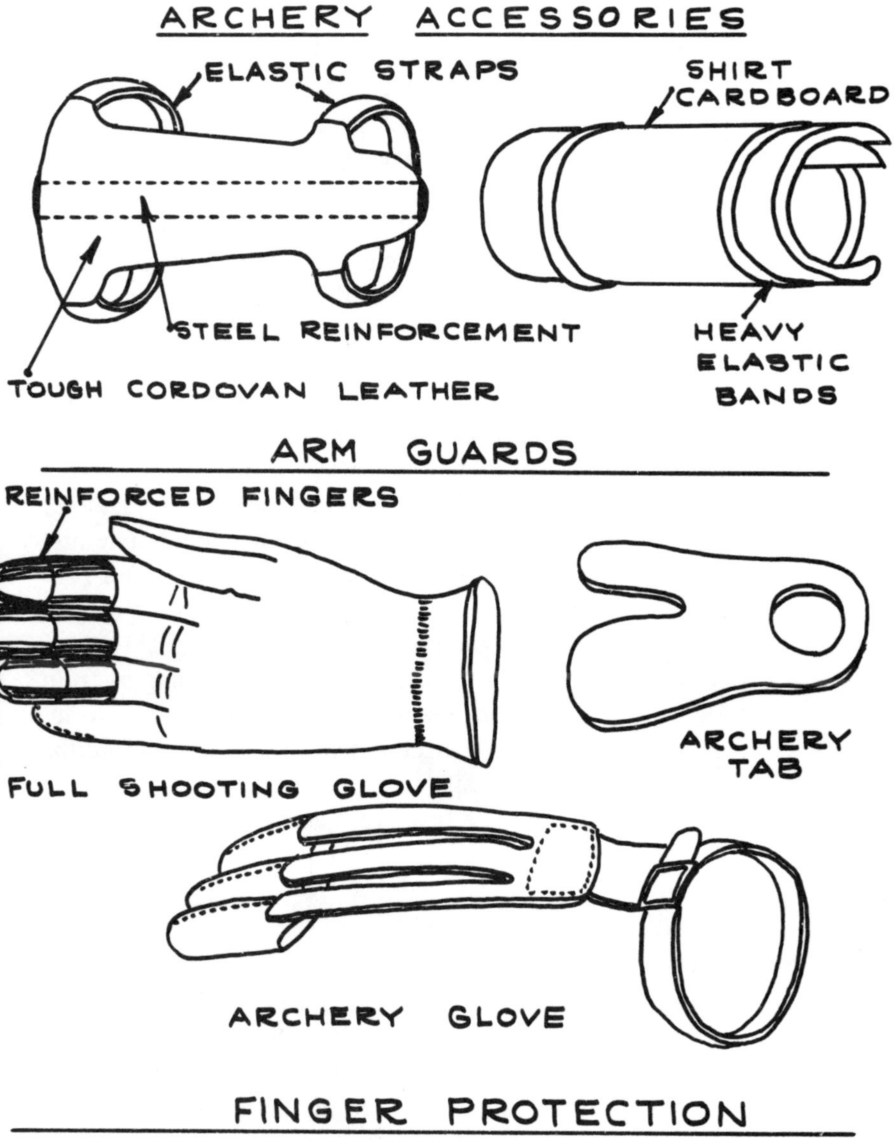

ARCHERY ACCESSORIES
ELASTIC STRAPS
SHIRT CARDBOARD
STEEL REINFORCEMENT
HEAVY ELASTIC BANDS
TOUGH CORDOVAN LEATHER
ARM GUARDS
REINFORCED FINGERS
ARCHERY TAB
FULL SHOOTING GLOVE
ARCHERY GLOVE
FINGER PROTECTION

FIGURE 1

A full shooting glove is shown in the middle left-hand illustration of *Figure 1*. These gloves are an excellent way for the beginning archer to get started, for the construction is familiar, and it is an easy transition from a regular glove. It is important that the finger reinforcement be quite stiff leather, or the bowstring will still bruise the fingers.

A three-fingered archery glove is shown in the lower illustration of *Figure 1*. This odd-looking piece of equipment is actually about the best finger protection available at the present time. These gloves are made with very stiff leather in the finger tips. The glove must be molded to your individual fingers by soaking the finger tips in water for ten minutes, and then molding in creases at the proper places for your finger joints. The glove is then allowed to dry out, and it will retain the proper shape for your hand. The first few times this glove is used you will drop arrows, and generally have a bad time. Once you get the hang of it, you will find this glove a great aid to good shooting. The finger tips remain in exactly the same place on your fingers from shot to shot, so that the release of the arrow is more uniform. The tough leather tips permit extended shooting before the fingers tire. An elastic strip between the leather fingers and the wrist band holds the tips in proper shooting position.

ARROWS

Most archers start purchasing a few arrows at a time from open racks in sporting goods stores. During early practice breakage and loss are frequent, and this approach allows a nominal set of ten arrows to be maintained by purchasing a few arrows at a time. Select the arrows for straightness by putting the point on one finger of the left hand, and sighting along the arrow while twirling it with two fingers of the right hand. A wide variation in these arrows exists, and only the straightest should be chosen.

As soon as possible, a matched set of wood or fiberglass arrows should be purchased. The arrows should be of proper stiffness to match your bow, and should be long enough for your draw. Standard 28-inch arrows are not long enough for many men, and advice from an experienced archer should be secured if you have questions on proper length.

The construction of arrows, and accuracy are discussed in Chapter 3.

QUIVER

The quiver is not an essential piece of equipment for target archery. The quiver swinging at the waist with long arrows protruding can be a real nuisance to the beginner. If the boxed set of arrows can be placed on a table or chair near the firing point, it will provide a convenient place to keep arrows, pencil and paper, and spare archery equipment.

A "ground quiver" is a handy device for target shooting. This consists of a short post stuck in the ground with a loop at the top 5 to 8 inches in diameter. The arrows can be placed through the loop with the points on the ground, and will be held vertically in a convenient position for use. The ground quiver must be made of fairly heavy wire, or the weight of the arrows will bend the vertical stake. A good material to use is 3/16 inch diameter iron rod, or cold rolled steel. The rod should be bent at a right angle about 18 inches from one end. The wire should then be bent to form a circle 8 inches in diameter parallel to the ground. The end of the 18-inch section of rod should be sharpened to a point. To use the ground quiver jam the sharpened end into the ground a short distance to one side of your firing position. The 8-inch loop forms a circle parallel to the ground and about 15 inches above the ground. Your dozen arrows can be dropped through the circle point down, and they will stand vertically in a very convenient position for shooting. Be sure that the ground quiver is not in front of your bow, or there is a good chance the lower bow limb will strike the quiver when you release an arrow.

A portable quiver is essential for field shooting and hunting. There are four basic types in use today, and many subdivisions of these types. The most widely used quivers are the traditional ones that hang from the belt, and hold arrows at a convenient angle to reach. These quivers are made in dozens of different shapes and sizes. A typical design is shown in the upper left-hand sketch in *Figure 2*. These quivers are usually slung fairly low for target shooting to make the arrows easy to reach. This position is not good for field use, as the quiver swings around too much. If the quiver is slung high, near the belt, then walking is easier, but the arrows are harder to withdraw. The arrows are relatively unprotected in this type of quiver, so that care must be used in walking through any type of brushy country.

The second type of quiver in wide use is the back quiver, shown in the upper right-hand sketch in *Figure 2*. These quivers also come in a wide variety of styles and price ranges. All are positioned on the archers back by some arrangement of straps that go over one or both of the archer's shoulders. The better models also have a cross strap of some type to prevent the quiver from swinging from side to side. The quiver shown in *Figure 2* has a strap going over each shoulder, and a cross strap that goes across the chest. The arrows stand vertically in the quiver, and are withdrawn by reaching a hand over the shoulder, grasping the neck of an arrow and lifting it out of the quiver vertically.

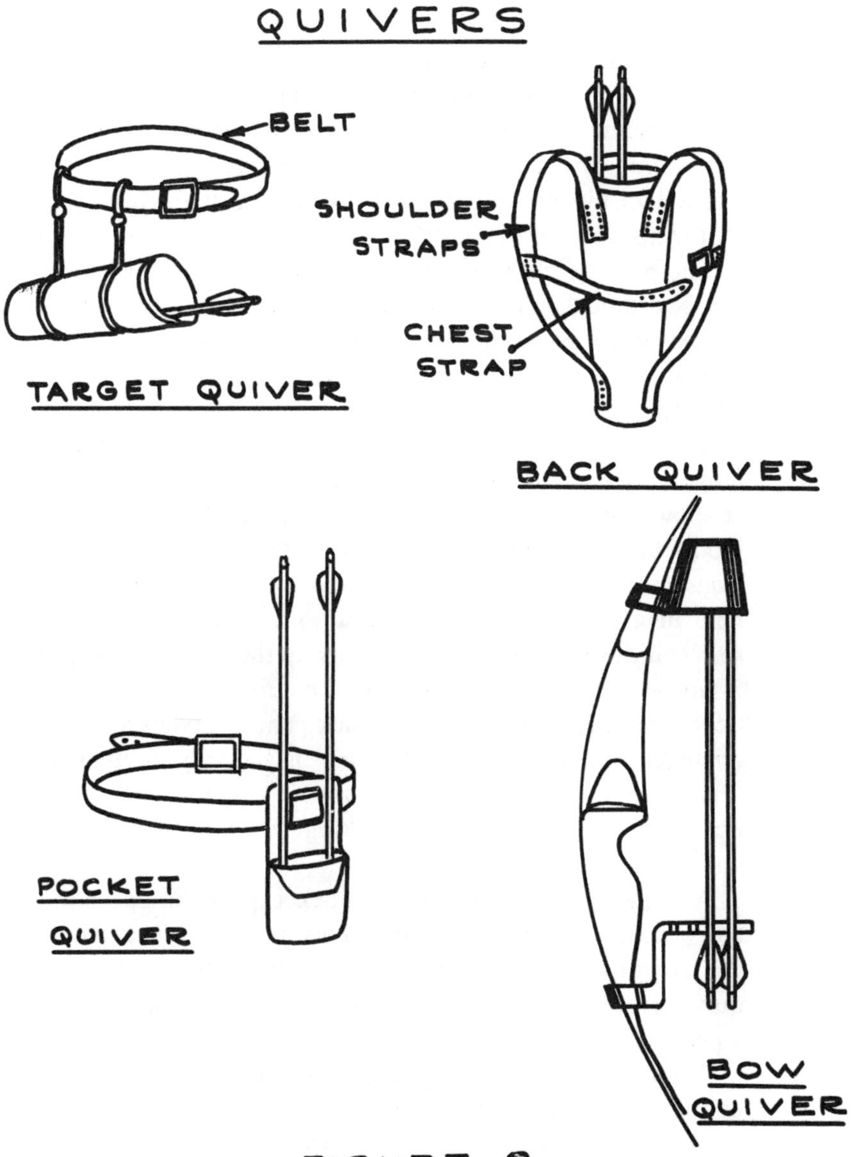
QUIVERS
BELT
TARGET QUIVER
SHOULDER
STRAPS
CHEST
STRAP
BACK QUIVER
POCKET
QUIVER
BOW
QUIVER

FIGURE 2

As soon as the arrow clears the quiver the archer swings his hand forward, pivoting the arrow through a wide arc and bringing it right into position to be placed on the bowstring.

A different type of back quiver is the "St Charles" model which holds the arrows vertically in a rack. The arrows are withdrawn by reaching around the back at waist level and removing them horizontally.

The back quiver is generally a better hunting quiver than the belt model. The hunter must be careful with a back quiver when going under low-hanging branches, but the quiver stays in place, and does not swing around while hiking.

The third type of quiver is a very compact "pocket quiver" which holds three or four hunting arrows. This type is shown in the lower left-hand sketch in *Figure 2*. This model fits on the hunter's belt and slides into the right rear trousers pocket. The lower ends of the arrows are firmly held in a point retainer at the bottom of the pocket, and small guides grip the arrow part way up the shaft. The upper 20 inches of arrow projects above the quiver and stands vertically, parallel to the shooter's back. These models are very compact and convenient, but are necessarily limited in arrow capacity. They are excellent for going through brush, but care must be used in working through low-hanging cover.

A larger quiver similar to the pocket quiver in operation has recently been introduced. This is a larger capacity model which holds the arrows in a similar manner, but the quiver is strapped to the hunter's leg rather than fitting into a pocket.

The fourth type of quiver is a rack which fastens directly to the bow. These quivers come in several styles, and are made from plastic, neoprene, or wire and sheet metal. The quivers generally are made with two racks which are spaced two feet apart, and hold the hunting arrows in two places. A typical bow quiver is shown in the lower right-hand sketch in *Figure 2*. The racks attach to the tapered ends of the bow riser, and are held on with tape, screws, or clamps.

This type of quiver is unorthodox, but it makes good sense. The bow riser section is rigid and does not flex during use, so that the quiver has little or no effect on the bow-operating characteristics. Placing the quiver on the bow frees the archer from any bulky quiver on his body. Care must be used in maneuvering any bow through brush, and the addition of a quiver to the bow does not add much to the problem.

New and improved quiver designs are being developed constantly, and this is another example of the dynamic new thinking going into improved archery equipment.

ARCHERY TARGETS

The selection of a suitable archery target is another choice that must be made. One of the most popular portable targets consists of layers of corrugated cardboard disks. These targets are available at relatively low cost, and are often used by beginning archers. A target 36 inches in diameter is recommended, since this provides a good size to aim at, and yet is not too heavy to handle.

After several hundred shots have been fired at the laminated cardboard target, arrows will begin to go through. There is an excellent means by which these targets may be repaired and made much more rugged. When the target can no longer satisfactorily stop arrows, disassemble it completely. The front cardboard disks will be perforated with a series of holes. The disks towards the back of the target will generally have large tears in them and must be discarded.

Sometimes excelsior is packed in between the disks and this is an excellent material for stopping arrows. This should be reused in the new target. Secure some large cardboard cartons, and cut some new corrugated disks using one of the original disks as a pattern. Secure some old rugs and cut circles of rug the size of the cardboard disks. Rug underpads may also be used for this purpose, but they do not have the toughness and wearing qualities of rug material. Four to eight layers of rug make an excellent pad for stopping arrows. For a bow over 40 pounds, at least six layers are required to stop arrows over a long period of time. A laminated sandwich consisting of a layer of cardboard, then two layers of rug, another layer or two of cardboard and excelsior, several more layers of rug, and several layers of cardboard at the back, make a very effective and tough target. It is well to stitch this sandwich together using heavy string. Holes may be driven through using a sharpened wire coat hanger or an ice pick. A new target face on the assembly is the last element to provide a tough and long-lasting target.

GRASS TARGETS

The best commercial targets are made of grass or straw compressed into strings which are then spiral wound around a wood excelsior core at the center of the target. These targets last much longer than a corrugated cardboard target. A 36-inch diameter target generally weighs about 35 pounds, and is an excellent all-around size. A 48-inch diameter target is much easier to hit since it has nearly twice the area of the 36-inch target, but the weight is up close to 60 pounds which makes it rather unwieldly for handling in the field.

Target arrows are deliberately designed with inefficient points so that they do not penetrate too far; but the new fiberglass and aluminum arrows are smaller in diameter than the wood arrows and, therefore, their penetration qualities even with relatively blunt points are fairly great. Heavy bows aggrevate the problem with any arrows. If arrows penetrate the grass target, an excellent solution is to fasten three or four layers of old rug to the front face of the target. This will toughen up the structure and make it much more resistant to arrow penetration.

TARGET STAND

A three-legged stand is a useful accessory to hold the target. The design of a simple stand suitable for use with any 36-inch diameter target is shown in *Figure 3*. The stand may be made from any scrap lumber such as one by two or one by three inch lath. A front view of the stand is shown in the upper left-hand illustration. First cut a 12 inch by 12 inch piece of ¼ or ⅜-inch thick plywood. Next cut a 16-inch-long vertical support, and two 36-inch-long legs. Fasten these securely to the plywood in the position shown in the drawing. Either nails or screws can be used, but the fastening should be thorough.

The rear leg should be cut from the same lumber, about 27 inches long, and bevelled at both ends as shown in the upper right-hand drawing in *Figure 3*. The rear leg is held on by two angle brackets each 2 inches by 2 inches long. One leg of each bracket is bolted to the plywood, and a single bolt is fastened through the other leg of the brackets, and through the rear leg of the target frame. A detail of this construction is shown in the side view of the frame in *Figure 3*.

The target rests on two large brackets which are fastened to the front legs. Ordinary stamped sheet metal shelf brack-

TARGETS

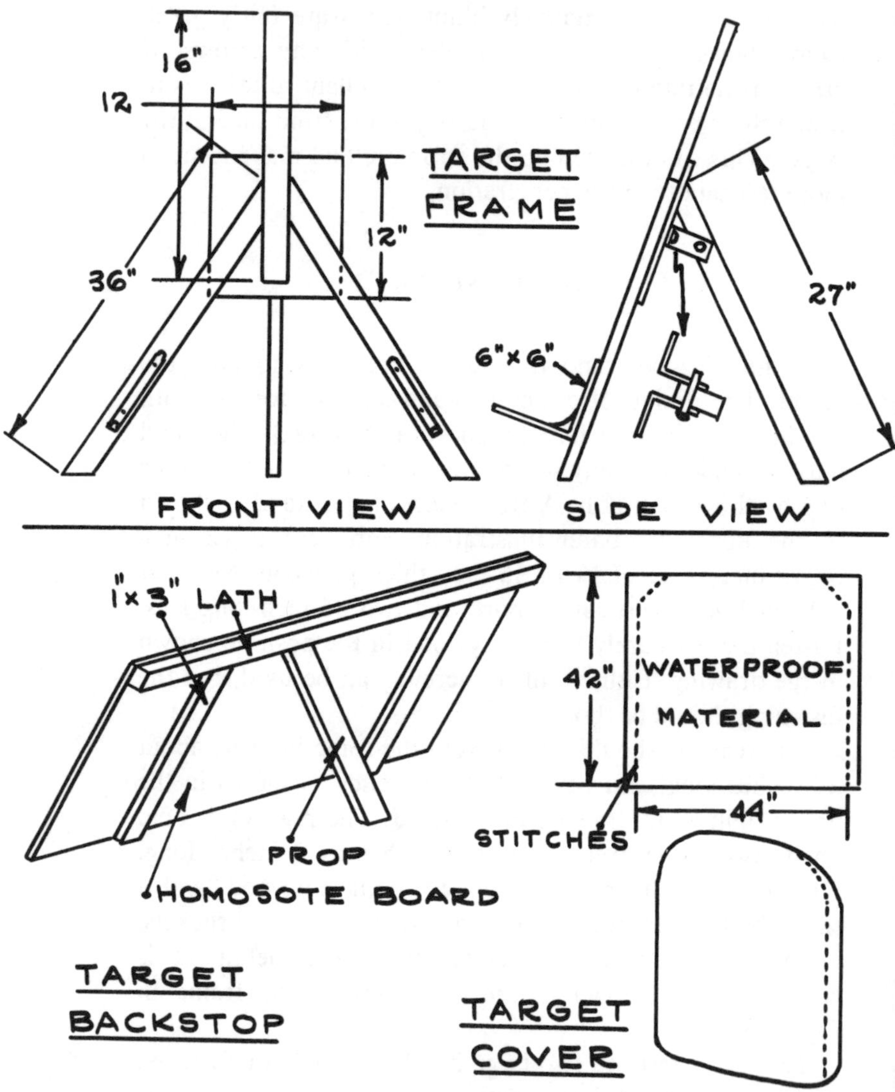

FIGURE 3

ets can be purchased at any hardware store, and these will do the job nicely. They are installed "upside down" as shown in the illustration. In this way an arrow cannot strike the vertical leg of the bracket, since it is protected by the full thickness of the target. The author originally constructed his stands with the brackets "right side up," leaving the vertical leg exposed. It is inevitable that arrows will strike these surfaces, and even a fiberglass arrow will be damaged if it hits a flat metal surface when it is travelling at full velocity.

The target stand must be angled backwards so that the center of gravity of the target is behind the front legs. If your target tends to fall over frontwards, shorten the rear leg to give more backward tilt to the frame.

The target and frame may be easily moved as a unit by standing in front of the target and reaching over and grasping the rear leg of the frame. The entire assembly can then be lifted as a unit and carried like a heavy suitcase. A 48-inch diameter target gives almost twice as much target area to shoot at, but these targets are quite bulky and require two men to move them around. One advantage to the 48-inch size is that it is the standard outdoor tournament target, and this is important if you are planning to enter competition.

The author uses two 36-inch diameter targets which can be set at two different ranges to give beginning shooters a handicap, and still allow side-by-side competition in the back yard. A range of 20 yards is a good distance at which to start. The range can gradually be increased as your skill improves. It is possible for an average archer to keep all of his arrows on a 36-inch diameter target at a range of 50 yards if there is no gusty wind blowing, and if he is using all of the aids available to modern archers, such as matched arrows, and some kind of bow sight. Modern archery equipment has superb accuracy potential, far beyond the capabilities of the average equipment available a generation ago.

TARGET COVER

A simple waterproof target cover will protect the target from rain and snow. A simple cover can be made in the shape of a large paper bag, and dropped over the target and frame when the day's shooting is over. A design for a target cover is shown in the lower right-hand side of *Figure 3*. Any waterproof plastic or cloth can be used, but do not use any of the very light-gauge plastics, which will tear easily and which can be lifted right off the target by a gust of wind. A piece of material 4 feet wide and 7 feet long can be folded over double with the fold at the top of the illustration. The sides may then be stitched together with coarse stitches using string or heavy thread. A heavy sailmaker's or shoemaker's needle is handy for this kind of sewing. No matter how horrible your sewing turns out, it can all be hidden by turning the bag inside out after it is finished. The end result looks as shown in the lower right-hand illustration in *Figure 3*. It is simple to make and gives excellent protection.

TARGET BACKSTOP

An excellent backstop for any archery target can be constructed at low cost. A four by eight sheet of "Homosote" wallboard can be purchased for only a few dollars. Several reinforcing boards should be nailed to the "Homosote" using regular large headed plaster nails. The face can be made waterproof with any type of paint. The "Homosote" panel can be placed several feet behind the main target, and held up near a vertical position with a prop placed behind, resting on one of the wood reinforcements. This large backstop catches 99 per cent of the arrows that miss the target, and shows immediately the direction of the miss. A drawing of the target backstop is shown in the lower left side of *Figure 3*.

BACKYARD ARCHERY RANGE

A backyard archery range may be constructed in many locations. The space requirements are not very great, and the land does not have to be level. One backyard range is shown in *Figure 4*. This range is not at all level with a drop of 15 feet between the firing point and target when the extreme range of 50 yards is used.

There are really only three essential elements required for an archery range. These are (1) a convenient shooting position with enough space so that spectators may stand safely to one side or behind the archer (2) a distance of at least 25 to 30 yards from the firing point to the target area (3) some means of catching arrows which miss the target.

The range shown in *Figure 4* solves these problems in one way, and there are many other equally good solutions. The Figure shows the author's wife using a bow sight on a light-weight target bow, with the target set at a range of 30 yards. The straw target has been reinforced with three layers of rug stitched to the target face. The target has been placed on a stand, such as that shown in *Figure 3*. In back of the target is a 4 foot by 8 foot sheet of fiber wallboard used as a backstop for the target. Well in back of the target is a genuine Connecticut stone wall dating from about 1830. This wall is composed of angular rocks which will chew up any kind of arrow that hits at a high velocity. Wood and aluminum arrows are inevitably broken or bent after an encounter with the wall, but fiberglass arrows usually have

Backyard range showing ground quiver, and target on frame with backstop behind target.

FIGURE 4

a chewed point which can be repaired. The Homosote backstop prevents 99 per cent of the misses from getting to the stone wall. The occasional arrow that goes through the Homosote has a greatly reduced velocity and lands in the grass between target and wall, a distance of about 20 yards. Arrows that go over the top of the backstop are on their way towards a serious collision.

A close-up illustration of the target area is shown in *Figure 5*. A completed target assembly on its stand is shown on the left side of the illustration. This particular target has seen a fair amount of use, and a protective layer of rug material on the front face has been shot to bits and replaced with new material. This was originally a laminated cardboard target which has been rebuilt with layers of rug, new cardboard, and rug underpads to make a much more rugged and permanent type of structure.

The target backstop is located immediately behind both targets. This stop was made from a sheet of "Homosote"

The construction of targets and stands is shown in this illustration. On the left is a target assembly on its stand. Behind the target is a backstop of fiber wallboard. A bare target stand is shown in the middle of the illustration. A partially disassembled target is at the right side of the illustration with a cloth target face on the ground in front.

FIGURE 5

wallboard, and it is scarred from numerous missed shots. It has remained outside all through the Connecticut winters, although a gusty wind broke one of the wood reinforcements. This was easily repaired, and the backstop was given a coat of paint to prepare it for another season of abuse. Arrows will occasionally go through, but they travel a very short distance and are found on the grass behind the backstop.

A bare stand is shown in the center of the illustration. This design is very similar to that shown in *Figure 3* except that the target-support brackets have been bent from heavy aluminum bar and bolted to the legs of the stand. After one fiberglass and several wood arrows sustained severe and permanent damage in head-on collisions with this bracket, large blocks of wood were fastened to cover the exposed aluminum. On the other target stand the brackets were turned upside down. A partially disassembled target rests against the backstop. This grass target would stop wood arrows very well, but aluminum and fiberglass arrows tended to drive through. Three layers of rug on the front stopped all arrows, but the center of the target took a great deal of wear and tear and eventually had to be replaced. In this illustration new rug material has been stitched in between the old cardboard sheets. The target is ready for reassembly with a new target face.

The power of arrows after a 50- or 70-yard flight from even a medium-powered bow is quite impressive. Before the rug material was added to the target aluminum arrows would frequently penetrate the target and drive through a one-quarter-inch thick piece of plywood, sticking so hard that they would have to be driven out with a hammer.

During archery sessions in the backyard the arrows are kept in a box which is laid across the arms of a lawn chair. Sometimes a ground quiver is used. A pencil and paper are kept handy to write down the range, wind conditions, sight settings, and scores. A wind toward the archer will drop the center of impact of your group, and a side wind of course moves the group in the direction of the wind. The sights can be set to compensate, once you get the hang of it.

At the end of an archery session the arrows should be examined for splits, loose points, or loose feathers. If the arrows are repaired immediately after the archery session the repairs will have dried, and a full set of arrows will be ready for the next day's shooting session. In between sessions the arrows are best stored in their box, or in a rack which prevents the feathers from being crushed. If the feathers do become a little battered, twirling them in the steam jet from a kettle will quickly restore their shape.

The author does not unstring his bow unless there will be no archery for three or four days. Leaving a bow strung can do no harm to a modern, well-constructed, laminated bow, but the string will gradually stretch. The greatest risk to a modern recurved bow occurs during the stringing and unstringing operations. Therefore it makes good sense to risk stretching a one-dollar bowstring, rather than risk twisting a bow limb by frequently stringing a seventy-five dollar bow. Modern Dacron strings are very tough, and last very well even if the bow is left strung.

The selection and care of proper archery equipment takes some thought. It is hoped that the information in this book will help you to make a better selection of equipment, and to use it with more skill and pleasure.

2
The Bow

The origin of the first bow is lost in man's early history. It was probably a logical invention after the spear, which had shown hunters the benefit of an instrument that would kill animals from a short distance.

The earliest bow was a flexible wooden sapling bent into a "D" shape with an animal sinew string. The arrow used was a straight stick sharpened at one end, with a simple notch at the other end.

By the Mesolithic Period (8,000 B.C.), the bow and arrow were in wide use in Northern Europe. The bows were made of water-softened ash, and animal sinew strings were used. Two types of arrow were in use. Flint-tipped arrows were used for big game, and arrows with blunt wooden heads were used for bird shooting. Prior to World War II, the vast majority of archery equipment used throughout the world was a logical development of these early systems. Bows made entirely of wood were used with wooden arrows. The arrow was drawn with three fingers of one hand, and it rested on the top of the other hand, which gripped the bow. Sighting systems were rare, and the accuracy in average hands was pretty poor.

After World War II, experimenters and far-sighted businessmen started applying the advanced technologies developed during the war to the field of archery. The result has been a complete revolution in archery equipment over the past 20 years. The all wooden bow has been almost completely replaced by a solid fiberglass bow in the cheaper archery equipment, or by laminated wood and fiberglass bows in the better quality levels. Arrows are now made of wood, fiberglass, and aluminum. While beginners almost invariably—and correctly—start with wood, the serious archer will usually switch to fiberglass arrows before many seasons are over. If tournament archery is the goal, then the more accurate, but more fragile, aluminum arrows are usually chosen.

A comparison of the construction of the old wooden bow and a modern fiberglass bow limb is shown in *Figure 1*. The wooden bow has limbs of a "D"-shape section. The proportions of the dimensions on the wooden bow have varied tremendously over the years, but the one shown in *Figure 1* is a typical construction. The flat section of the bow faced the target. This section was placed in tension as the bow was drawn. The rounded section or "belly" of the bow faced the archer and this section was placed in compression as the bow was drawn.

RECURVED BOW DESIGN

The design of a modern laminated bow differs radically from the older bow in other ways. The old wooden bow was straight when unstrung. It was held in an arched form by its string. There was generally a distance of about 7 inches between the string and the bow with the bow properly strung. As the bow was drawn, the curvature of the limbs

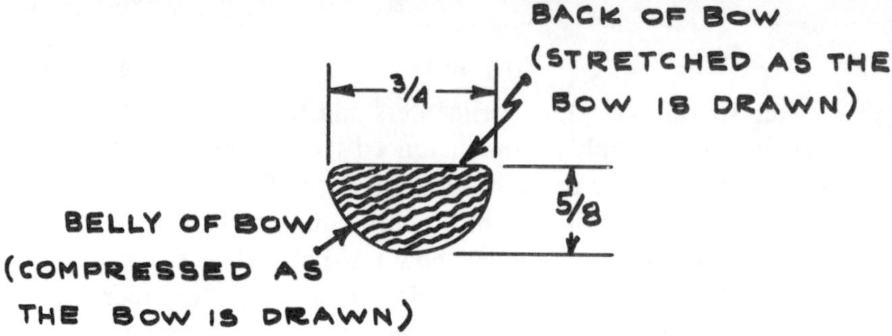

CROSS SECTION OF A WOOD BOW LIMB

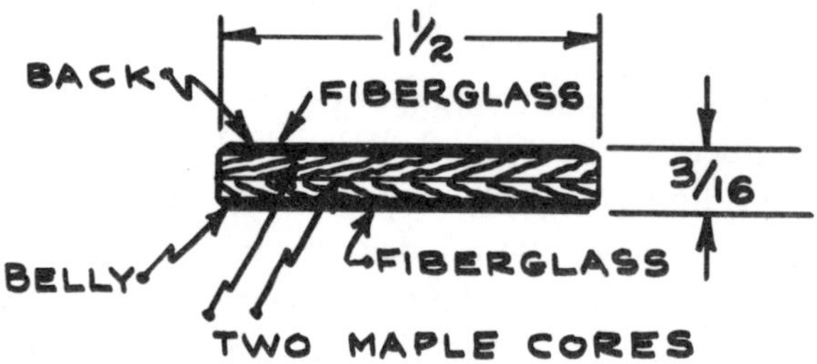

CROSS SECTION OF A LAMINATED LIMB

FIGURE 1

became more acute and all of the bow except for the handle section deflected as the drawing force was applied. In effect, the bow was composed of a short center handle with a 30-inch cantilever beam extending from each side. As the arrow was released, the entire upper and lower bow limbs moved forward to return to the rest position.

The construction of the modern laminated bow is entirely different. The bow is shown in the rest position on the left side of *Figure 2*. The center 28 inches of a first-class modern bow are rigid and do not deflect at all as the bow is drawn. The entire deflection occurs within the 18-inch upper and lower limbs. In the rest position, the upper and lower limbs are curved sharply forward. The bow properly strung is shown in the center of the illustration. The shape of the upper and lower limbs is radically changed. The fiberglass on the face of the bow has been placed under considerable tension. The extreme tips of the limbs are still angled forward far enough so that the string lies along the surface of the bow limb for a distance of 3½ inches. As the bow is fully drawn, the upper limbs unwind even further so that they form a graceful "S" curve. When the arrow is released, only the upper 18 inches of each limb moves and only the outer 10 inches moves very fast.

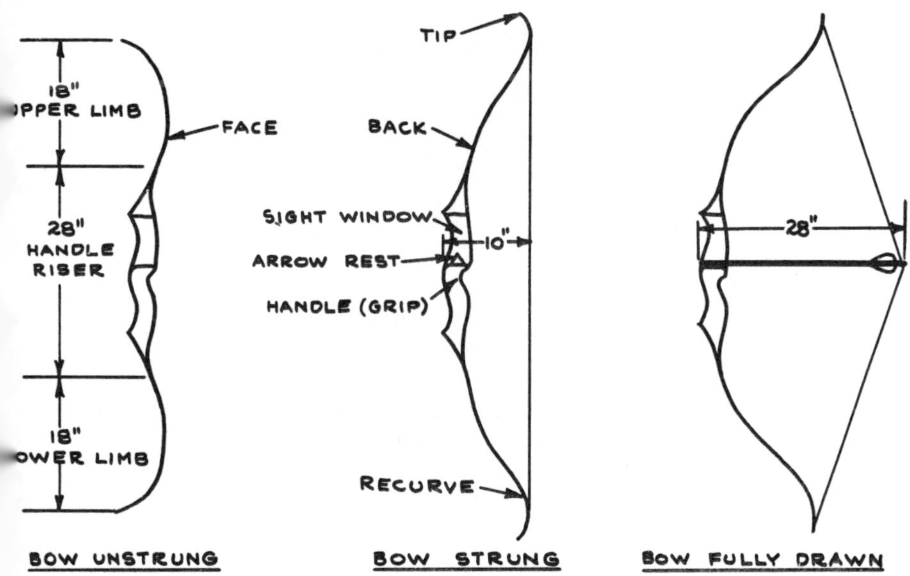

FIGURE 2

The overall result of this new design is that a small mass of bow limb moves as the arrow is released, and a higher percentage of the system energy goes into accelerating the arrow. Thus the spring characteristics of the system are improved and the arrow velocity considerably increased.

There are many different designs of bows to meet the needs of today's archers. While some all-wood bows are

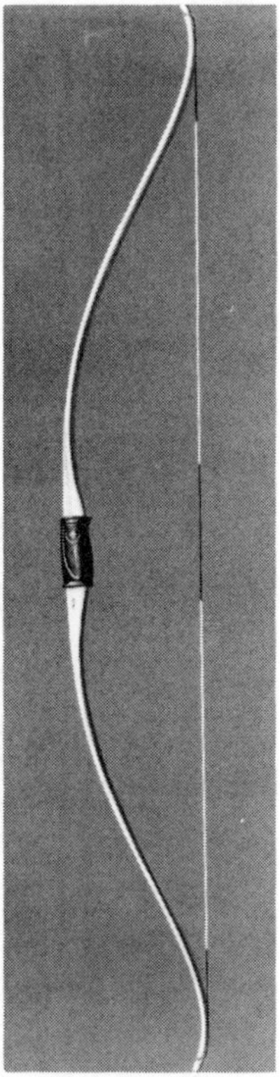

Ben Pearson Model 3350-0 solid fiberglass bow.

FIGURE 3

still manufactured, major manufacturers have switched to solid fiberglass bows in this price range. A Ben Pearson Model 3350-0 solid fiberglass bow is shown in *Figure 3*. This bow is known as a semi-recurved design since the bow tips are not arched as far forward as on the fully recurved bows. Overall length of the bow is 60 inches and it is available in draw weights from 25 to 50 pounds. Normally a bow length of at least 60 inches should be selected for a general-purpose bow for adults. A plastic handle is molded in the center to provide a comfortable grip. The upper edge of the handle provides a shelf on which to rest the arrow, and the bow is designed symmetrically so that it may be used either right or left handed. The center handle is injection molded of heavy duty vinyl plastic and is textured to provide a comfortable but nonskid grip to assist in holding the bow accurately. The bow is fitted with a dacron bowstring consisting of many fine strands of dacron thread. The ends of the string and the center are wrapped with thread to provide reinforcement at the main wear points.

It is a good idea even with a relatively inexpensive solid fiberglass bow to put a "nocking-point" on the string at a location which will make the arrow exactly perpendicular to the string when resting on the handle. This can be done by taking some white thread and wrapping it around the bowstring half a dozen times, then tying knots and putting on a touch of fletching cement. It can also be accomplished using specially manufactured accessories which consist of a small plastic tube which is heat shrunk into proper place on the bowstring. Other manufacturers provide small rubber rings which may be attached to the bowstring to provide the proper nocking position.

The addition of a nocking-point is rather important even with a relatively inexpensive bow, for it insures that the

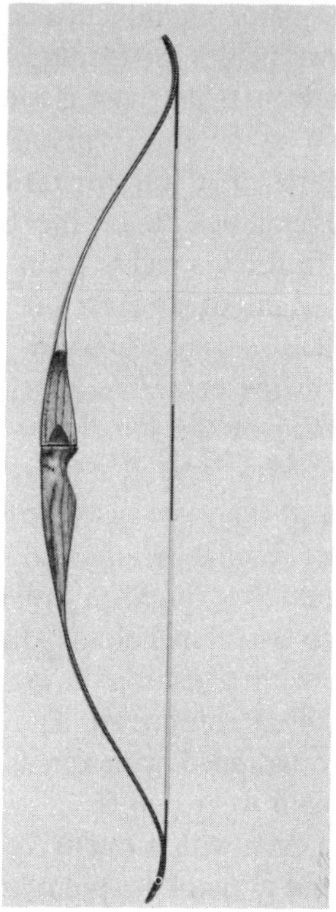

Pearson moderately priced wood and fiberglass bow (Model 7120-0 Varsity).

FIGURE 4

arrow is in the proper vertical position shot after shot. Solid fiberglass bows such as that shown in *Figure 3* are extremely durable and capable of good accuracy if used with care and practice. A good solid fiberglass bow suitable for use by an adult archer will range from $15.00 to $25.00 in price.

MODERATELY PRICED LAMINATED BOWS

Laminated wood and fiberglass bows are available over an enormous price range. Special models designed only for small children are available in the $20.00 price range. A 60-inch general purpose laminated bow suitable for use by children and adults is shown in *Figure 4*. This is a Pearson model 7120 "Varsity" bow with an overall length of 60 inches and available in draw weights of 25 to 45 pounds. This style of bow is available from all the leading archery manufacturers and is generally priced in the $25 to $50 range. For adult use, a bow of at least 60-inch length should be selected unless it is specifically designed for hunting. Hunting bows are generally shorter, ranging from 52 to 56 inches in length, with heavier draw weights and are slightly more expensive than the corresponding general purpose bows.

BOW FISHING

Hunting fish with bow and arrow is an interesting sport that often utilizes solid fiberglass, or moderately priced laminated bows, since the equipment is exposed to pretty rough conditions. All of the elements required for bow fishing are shown on a deluxe Bear bow in *Figure 5*. An 11-inch diameter reel is attached to a shaft on the front of the bow. It is designed much like a spinning reel used in fishing so that the line may unwind very rapidly with little drag. The end of the line passes through a cross hole in the arrow and is then attached to the arrow tip. The arrows are normally fitted with sharp pointed tips with spring barbs which will fold flat as the arrow penetrates the fish and then snap outward and lock firmly to

Bear Model 3000 bowfishing rig.

FIGURE 5

prevent escape. Since water ruins regular feathers very quickly, fishing arrows are fitted with molded rubber fletching as shown in the illustration.

Bow fishing is a relatively short range but exciting sport. Great care must be taken to stalk the fish and the refraction of the water makes aiming rather tricky. When the arrow strikes there is plenty of action and excitement before the fish can be landed.

HUNTING BOWS

Hunting bows are available in a wide range of styles and prices from many manufacturers. One of the most deluxe and compact hunting bows is the Bear super magnum 48 shown in *Figure 6*. This represents about the minimum practical bow length. The great majority of hunting bows fall between 52 and 60 inches in length. The Bear Super-Magnum 48 has been designed for the experienced hunter who wants minimum size for moving through heavy brush and under difficult stalking conditions. The construction is of highest quality with three extra plies of fiberglass and resin on the back of the handle for added strength and extra layers of fiberglass at the bow tips to reinforce these points of strain.

Hunting bows are generally quite difficult for the inexperienced shooter to handle. The Super-Magnum is available in draw weights from 35 to 70 pounds. Stringing a short bow with heavy draw weight is extremely difficult. With the short limbs there is quite an angle to the string when fully drawn, tending to crowd the fingers in toward the arrow more than with a long-limbed target-type bow. Hunting bows are special equipment designed for a few shots with heavy arrows under high stress conditions.

There is a natural tendency to consider purchasing one

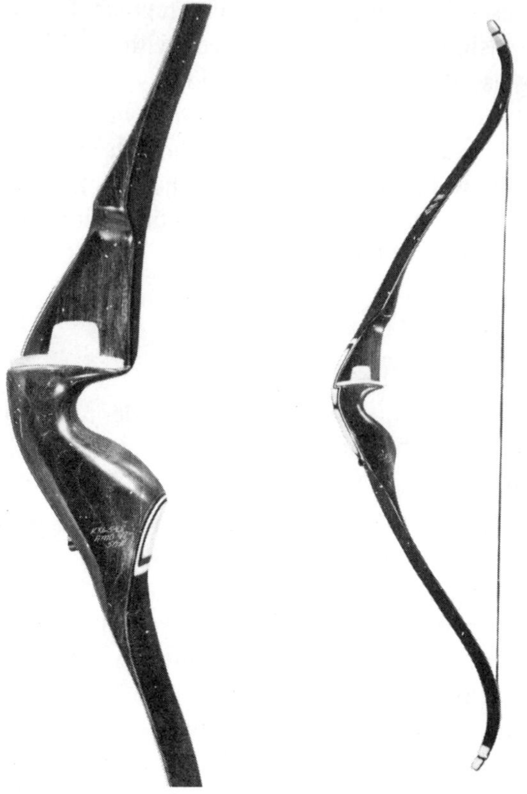

Bear Super-Magnum 48-inch hunting bow.

FIGURE 6

bow for use in target shooting and hunting. Generally this is not a good idea; for the average archer, skill should be developed using a general purpose or target-type bow rated between 30 and 35 pounds in draw weight. It is surprising to most people that tournament archers who are professional in every sense will normally concentrate in this range as providing the best opportunity to develop technique and high accuracy. A moderately weighted bow allows you to concentrate on achieving a full draw, holding the draw until the sight picture is correct and releasing carefully.

A very heavy bow requires intense muscular concentration and often leads to tremors and early fatigue for the average archer. It is much more difficult to concentrate on holding a full draw and developing a proper sight picture with a 50-pound bow than a 30-pound one, and the heavy bow often leads to discouragement and poor performance. A much better plan is to practice with a moderately weighted bow until technique is refined and then switch to the hunting bow for intensive practice and toning up of muscles just before hunting season. With this approach you can often carry target shooting skills and accuracy out into the hunting field.

TARGET BOWS

Target bows are available from many manufacturers in a wide range of designs and prices. Since extreme care is used in manufacture to achieve excellent balance between the limbs and very careful control of the draw characteristics, target bows tend to be rather expensive. One of the top quality full target bows is the Hoyt Pro-Medalist Bow shown in *Figure 7*. Another top quality target bow is the Pearson Centurion shown in *Figure 8*. Both of these bows combine beautiful laminated woods in the riser with extreme care and quality in the construction of the limbs. They can both be fitted with strange-looking devices known as "stabilizers." These bows are typically available in lengths of 66 and 70 inches and draw weights of 30 to 44 pounds. Such target bows are

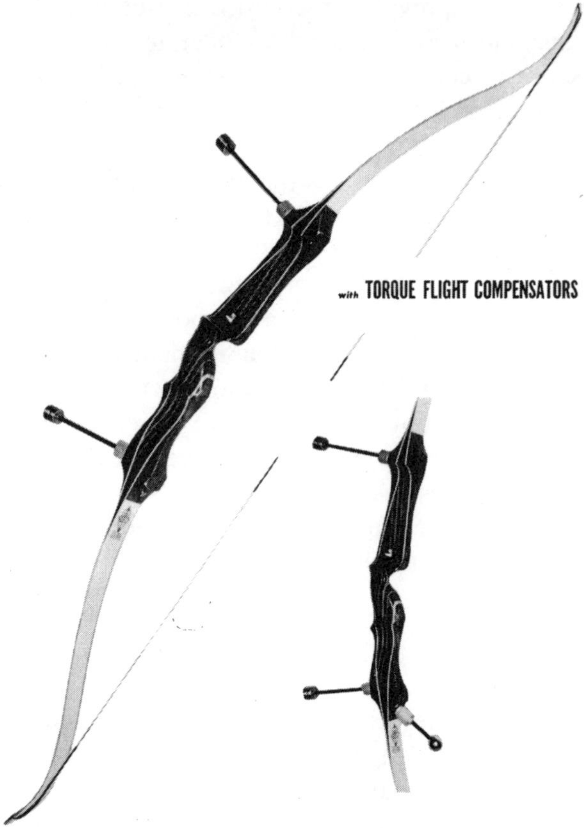

Hoyt Pro-Medalist target bow with compensators.

FIGURE 7

often constructed of such exotic woods as holly, cherry, bubinga or East India rosewoods. Some new target bows are manufactured with magnesium or aluminum risers, eliminating all wood except the laminates in the limbs.

The purpose of the stabilizers is very simple, and they

have proven to be quite effective for the extremely proficient archer. As the bow is held at arms length it is relatively easy to twist it in a lateral direction. During the release of the arrow and as the arrow snakes around the sight plate there are small lateral forces given to the bow and string which cause twisting. The purpose of the stabilizers is to minimize the twisting of the bow due to

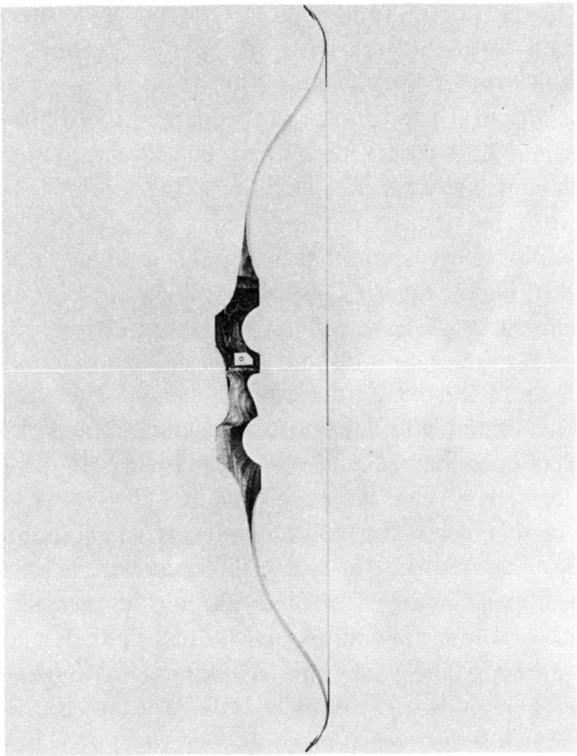

Pearson Centurion target bow.

FIGURE 8

these forces. In engineering terms stabilizers increase the longitudinal moment of inertia so that the bow moves much less for a given force. Most of the compensators are quite exotic in construction. For example, the Pearson bow has a sealed cartridge of liquid mercury contained in the metal barrel at the end of the stabilizer. Having a cavity partially filled with an extremely heavy liquid such as mercury is very useful in "damping out" the oscillations of the bow. This means that if the shooter has a poor release, the lateral force that his fingers give the string will tend to go into sloshing the mercury around inside the cavity rather than twisting the bow.

The Pearson Centurion has been redesigned for 1973 to include fine points developed in the recent years of tournament archery. Technically called a "deflex-reflex recurve," the notable feature of this type of design is a nearly flat reflex section between the working limb and the working recurve. Careful experiments on modern tournament bows have refined the design at the recurves to minimize the vibration and jar as the arrow is released.

On the Pearson Centurion the handle riser is made from laminated marblewood and shedua. The sight window has been moved, cut ⅜ of an inch past center, so that the arrow can be positioned at the center of the bow if this gives the tournament archer optimum accuracy.

The specifications for the Centurion provide two lengths—64 inch for juniors and 69 inch for adult tournament archers. The adult bow is available in weights from 30 to 45 pounds at a 28-inch draw. The brace height of these bows is held very close at 7½ to 7⅝ inches, and the other specifications are meticulously drawn for all of the top-quality tournament bows where a difference of one or two points may decide a match. Tournament archery has become so super-competitive at the national level that shooters are demanding, and getting, close to perfection in their equipment.

TAKEDOWN BOWS

The newest development in archery equipment is the takedown bows, which have been developed by some of the major manufacturers. The system developed by the Bear Archery Company is shown in *Figures 9 and 10.* The particular model illustrated is the Victor Magnum, which utilizes a handle riser cast and machined from magnesium metal. Magnesium is one of the lightest structural metals available and is used for certain components in high performance aircraft. It is considerably lighter than aluminum. The Victor Magnum takedown bow is available in draw weights of 25 to 70 pounds and the length may be varied from 56 to 60 inches by the use of different bow limbs. The hand grip is molded of black vinyl and is available in three different styles to fit the shooter's hands. The black plate in the sight window may be removed and a wide range of hunting or target sights installed.

The bow limbs are relatively conventional construction consisting of a wood core with fiberglass laminates on the front and back. Grooves have been machined into the limbs to provide exact alignment with the handle riser. The metal yoke and latch attached to the top and bottom of the handle riser are swung into an open position and the limbs slid into place. The design is such that the limbs can interlock with the handle in only one correct position. After positioning, the yoke is swung shut and

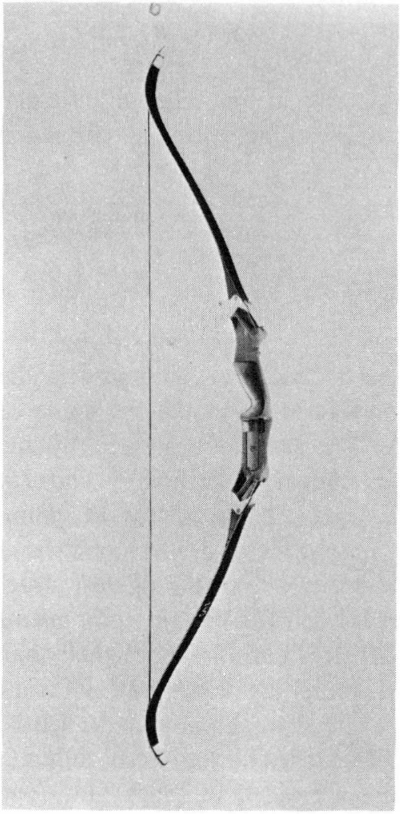

Bear Victor Magnum takedown bow.

FIGURE 9

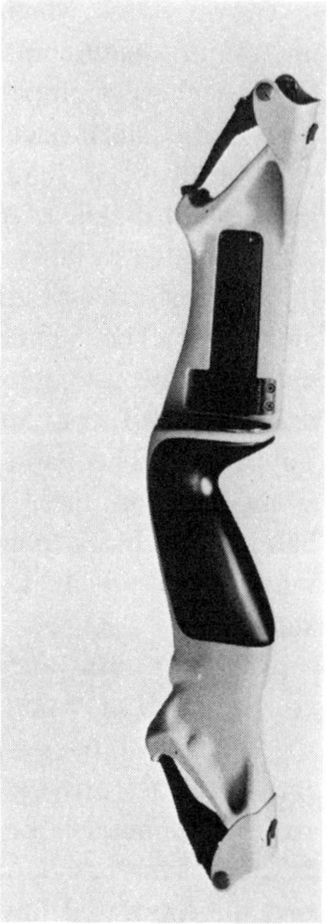

Bear Victor Magnum takedown bow: handle riser of magnesium.

FIGURE 10

latched, clamping the limb rigidly in place. The bow can then be strung and operated in the normal manner.

The Bear Company manufactures takedown bows for hunting, target, and general purpose uses. The magnesium risers are available in blue, gold, green, gray, red, brown, and rust color, while the wooden risers are manufactured in laminates of exotic and contrasting woods. Since such complex bows require a great deal of care in manufacturing, all are priced over $100.

A very different approach to the takedown bow has been developed by the Ben Pearson Company. An illustration of the Mercury Marauder takedown bow is shown in *Figures 11 and 12*. The Pearson bow divides into two assemblies and alignment is provided through the aluminum knuckle joint shown in *Figure 12*. The holes through the knuckle joint are tapered and the addition of a long tapered pin, which is threaded through the holes, provides extreme rigidity of the assembly. The handle riser is manufactured of a beautiful East India rosewood and provides a rich coloration. Overall length of the bow is 60 inches and it is available in draw weights of 40 to 55 pounds. Other Pearson takedown models, such as the "Competitor TD," are designed for target shooting and are available in lengths of 66 inches with draw weights from 30 to 55 pounds. Again, the handle risers are made of exotic imported woods such as bubinga and rosewood.

Takedown bows are very practical for the traveling archer. The takedown is quite simple; for example, on the Pearson bow you merely remove the string, unscrew the tapered pin and divide the bow in half. A soft vinyl case is provided that takes the two halves of the bow and provides a compact package, easy to carry and store.

The changes that have occurred in American archery equipment since World War II are amazing. The new

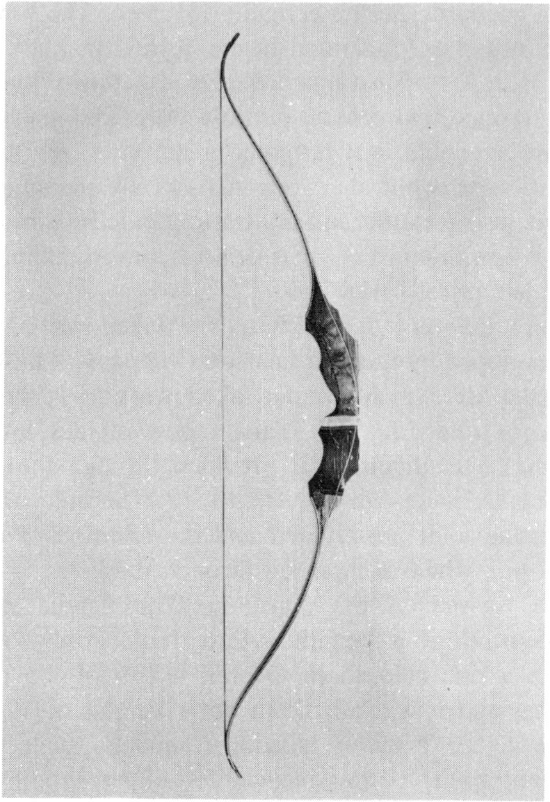

Pearson Mercury Marauder takedown bow.

FIGURE 11

full-target bows with carefully engineered draw characteristics, superb manufacturing, and stabilizers to increase accuracy are the finest target equipment in the world. The recent trend to takedown bow designs represents another level of sophistication for American equipment. Complex design and manufacturing problems had to be solved as

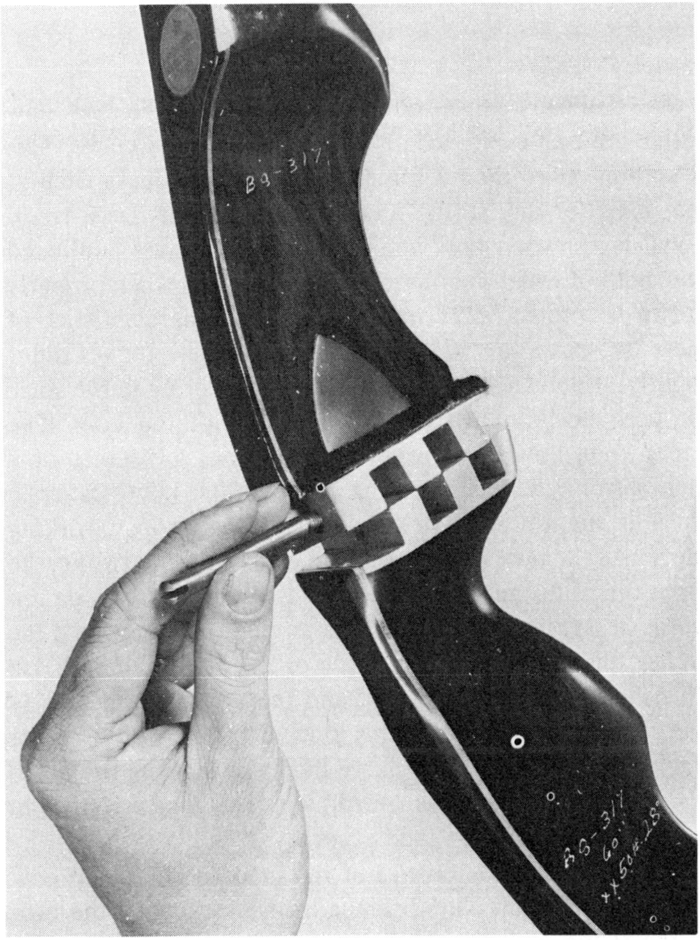

Detail of aluminum takedown joint on the Pearson Mercury Marauder hunting bow.

FIGURE 12

part of the continuing research and development of uniquely American designs.

FORCE VERSUS DRAW CURVES

The radically improved characteristics of the new bow design are best explained on a force versus draw curve. This is nothing more than a force versus distance curve such as that used for any spring analysis. The author has a traditional lemonwood bow and a modern fiberglass laminated bow both of which require exactly 36 pounds when properly drawn to project a 28-inch arrow. The characteristics of these two bows are shown in *Figure 13*. These curves differ slightly in nomenclature from a common spring curve since the zero position is taken as the front of the bow. The lemonwood bow when properly strung has a distance of 7 inches from the string to the front of the bow. Its draw curve is slightly concave upwards over its proper working range. As the bow reaches the end of its working stroke, the force versus distance characteristic is beginning to slope upwards at a faster rate, and this is termed "stacking." To the archer, this means that the draw of the bow is smooth, up till near the end of the draw, and then the force begins to rise very sharply. While this bow does not show severe stacking, the shape of the curve indicates that the force will rise rapidly if the bow is overdrawn. This characteristic in a bow is detrimental to good accuracy.

The spring characteristics of the modern laminated bow are quite different. The distance from the front of the bow to the string in the rest position is almost 10 inches. The spring force increases rapidly during the early part of the draw and the increase of force per inch of draw at the end of the draw is considerably less than with the old-fashioned bow. This has two beneficial results. First, the area under the curve for the modern bow is considerably greater than for the old-fashioned bow, thus more energy is available to

FORCE VERSUS DISTANCE CURVES

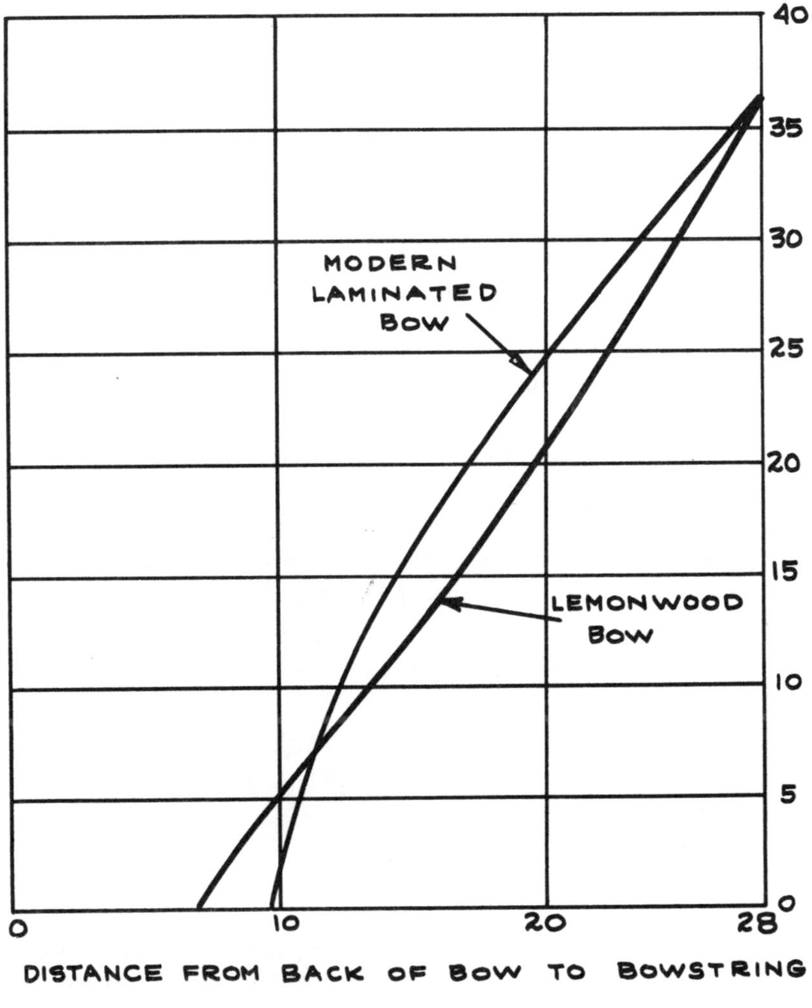

FIGURE 13

cast the arrow. The second benefit is that the bow will handle more smoothly to the archer. There is no abrupt increase in force as the end of the working stroke is reached and this modern bow may comfortably be drawn 30 or 31 inches in the hands of a long-armed archer.

The spring characteristics of three modern laminated bows of different force levels are shown in *Figure 14*. The lowest curve is for a very lightweight bow, suitable for use by women and youngsters. This bow is an excellent quality bow, so that even though the pull is light, the bow may safely be drawn 30 inches. This bow has delicate limbs, with thin maple cores and thin layers of fiberglass on the surfaces. The bow limb is only three-fourths as thick as that on the 36-pound bow shown in *Figure 1*. All the bow limbs taper in thickness from a heavy section at the center of the bow—known as the riser section—out to the tips of the limbs.

Light arrows should be used with this bow. Wooden target arrows properly spined for this bow weigh about 300 grains each (0.7 ounces). A comparison of the launch velocity of light target arrows and one-ounce field arrows is given below:

Arrow	*Arrow Weight*	*Velocity*
Light Target	300 grains (.7 oz)	147 feet/second
Field Arrow	437 grains (1 oz)	123 feet/second

The velocity of the heavy arrow is getting quite low, and the drop over a 50-yard distance would be almost 30 feet. This is an excellent bow for use at moderate ranges and with light arrows. Fundamentals of drawing, holding, and releasing the arrow can be learned much more easily with a light bow than with a heavy one. A good quality light bow, such as this, is an excellent first choice in a family with young archers.

The middle curve in *Figure 14* shows the force required to draw a modern good quality target bow. The bow is 66 inches long and requires 36 pounds of force to reach a draw of 28 inches. The bow may be safely drawn over 30 inches, and it is still smooth at a draw of 31 inches which the author uses with one set of wood field arrows. The middle curve in *Figure 14* is for the same bow as that shown in *Figure 13*. In *Figure 14*, the draw curve has been extended to 30 inches, where the bow requires 41 pounds of muscular pull.

Although this bow requires 36 pounds to reach a draw length of 28 inches, this maximum force varies greatly with draw length. Thus, a short-armed archer who draws 26 inches needs about a 32-pound pull. The normal archer needs 36 pounds at 28 inches. The long-armed archer needs to apply over 41 pounds of force to draw the bow for a 30-inch arrow. This again emphasizes the need to choose equipment on the light side, particularly if you have a long draw. The velocity of a light target arrow and a field arrow launched from this bow at a 28-inch draw are:

Arrow	*Arrow Weight*	*Velocity*
Light Target	300 grains (.7 oz)	189 feet/second
Field Arrow	437 grains (1 oz)	158 feet/second

Over a 50-yard range the one-ounce arrow would drop about 17 feet when launched from this bow. This is an acceptable drop, and accurate shooting can be done. The time of flight for a one-ounce arrow would be about one second over a 50-yard range compared to over 1⅓ seconds for the same arrow from the light 23-pound bow shown in the lowest curve of *Figure 14*.

The top curve in *Figure14* is for a heavier hunting bow with a rated draw of 46 pounds at 28 inches. This bow is more compact than a target bow, with an over-all length of 60 inches. Hunting bows generally are designed for a heavier draw than target bows. Many states require a bow of over 40 pounds draw for use in hunting, and the heavier pull is not so important out in the field where relatively few arrows per day are shot.

For target archery, bow weights under 40 pounds are generally recommended for all archers. Even the top-notch professionals usually choose a bow in the 30- to 40-pound range for tournament shooting. The longer limbs of a target bow give a smoother cast to the arrow, and reduce the vibration in the bow as the arrow is released.

The hunting bow shown in the top curve is designed for use with heavier arrows, such as one-ounce field arrows, or heavy hunting arrows of up to 1⅓ ounces. This bow would be very hard on light target arrows, but they are listed in the chart to show the velocity increase compared to the other bows in *Figure14*. The launching velocities of three types of arrows from this bow are:

Arrow	*Arrow Weight*	*Approximate Velocity*
Light Target	300 grains (.7 oz)	215 feet/second
Field Arrow	437 grains (1 oz)	172 feet/second
Heavy Hunting Arrow	600 grains (1.37 oz)	145 feet/second

The heavy bow gives excellent drive to the arrow, and reduces the time of flight and drop. With a one-ounce arrow at a 50-yard range, the time of flight is about .95 seconds, and the drop is only a little over 14 feet. The drop with this same arrow launched from the 23-pound bow was over 30 feet.

FORCE VERSUS DRAW CURVES FOR THREE LAMINATED BOWS

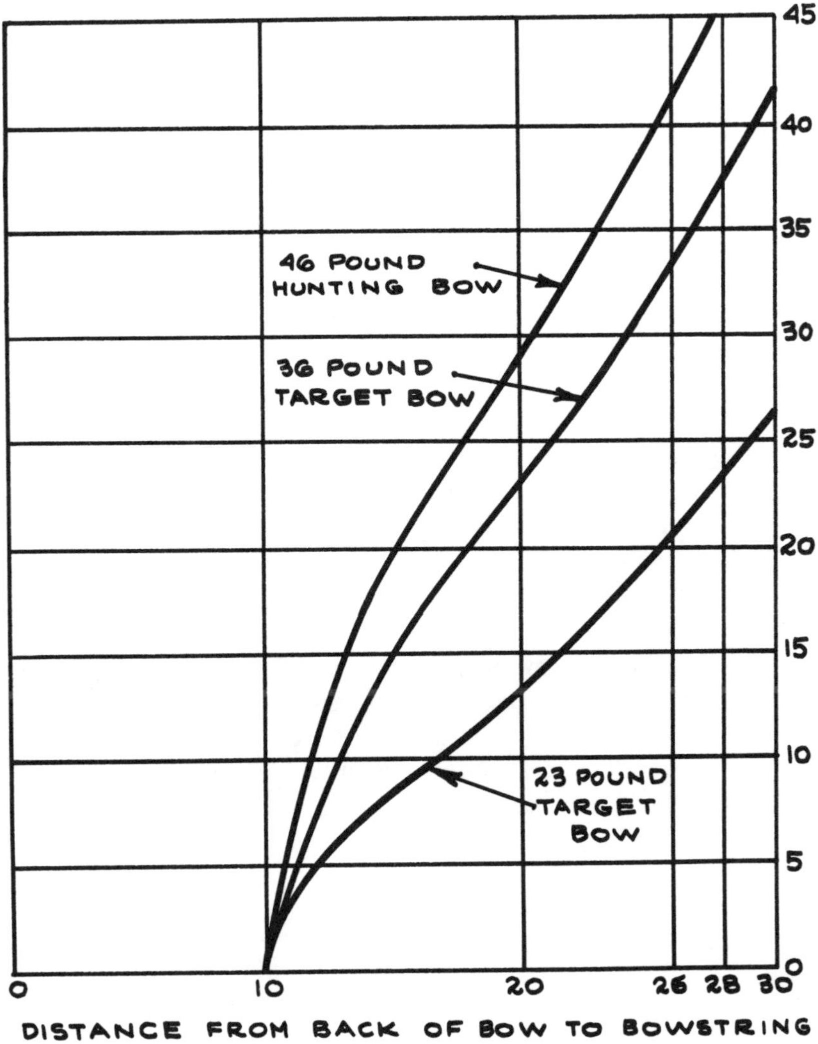

FIGURE 14

EFFECT OF DRAW LENGTH ON VELOCITY

Every archer draws his bow with his own style. The length of draw, type of release, and weight of arrow all affect the launch velocity of the arrow. The variations from one bow to the next due to differences in design and manufacture also affect launch velocity, even for bows of the same draw weight.

Of all the factors, the draw length has the largest effect for any given bow. This effect can be understood by referring to the force versus draw curves in *Figure 14.* The energy the archer must apply to the system in order to fully draw the arrow is shown by the *area under the curve.* For the 36-pound bow shown in the center curve the area represents 31 foot-pounds of energy at 28 inches of draw. This is equivalent to lifting 31 pounds one foot off the ground, or lifting one pound 31 feet in the air. In the case of this bow, the force is applied over a distance of 18 inches and the force required builds up from zero to 36 pounds.

The energy that actually goes to accelerate the arrow is less than the input. This is because of the energy required to accelerate the bow limbs, and the string, and the friction losses within the stretched fiberglass in the bow limbs. The output energy that the arrow receives amounts to about three-quarters of the energy input with top quality modern equipment. This does not apply to all wood or all fiberglass bows, but only to relatively expensive laminated bows which have been designed for efficiency and manufactured with care. The characteristics of this 36-pound bow over a range of draw lengths are given below:

Draw Length Inches	*Area Under Draw Curve (Energy Input in Foot-Pounds)*	*Approximate Energy Output (Foot-Pounds)*	*Approximate Velocity of one ounce Arrow Feet/Second*
26	25.4	19.7	143
28	31.2	24.2	158
30	37.6	29.2	174
31	41.1	32	181

Most beginning archers have a short draw length at the beginning of their career. As they develop proficiency, the draw increases slightly. One of the biggest problems in inexpensive equipment is a draw limit of 28 inches, which is not enough for many men. These bows tend to "stack" as the draw limit is exceeded, and may break due to overstressing if their rated draw length is constantly exceeded. It is much better to buy a good quality bow that is blemished, or of a discontinued style, rather than an inexpensive bow which will not give an adequate draw as the archer's shooting techniques develop. An understanding of the factors affecting performance in modern archery equipment can help in selecting proper tackle, and can help in using the equipment to best advantage.

3
Distance Shooting with Arrows

Over normal hunting and target-shooting distances, the drop of a rifle bullet is measured in inches. The elevation in angle of the barrel above the target seldom exceeds two degrees. Archery systems are used over much wider angles of launch. A standard archery game called "clout shooting" involves shooting arrows at elevations up to 45° above the horizontal, at a large target located 120 to 180 yards away.

Using a bow and arrow for extreme range shooting is very interesting, as it simulates the extreme range ballistics of rifles or artillery on a very small scale. If you are careful to launch the arrows at the same angle and length of draw for each shot, the accuracy will be very good in calm air.

With most bow and arrow systems, a range of 250 yards is long enough to carry out flight shooting. A deserted beach or level field are ideal locations. It is important to accurately measure distances out to the impact area using a 100-foot tape or other measuring system. Flight shooting should be done on a calm day, as the wind has a very marked effect over the long flight of the arrow. The ballistics of projectiles flying through space would be relatively simple if there was no air resistance. An arrow or bullet would travel along a parabolic arc, and would impact on a target (located at the same elevation as the shooter) with exactly the same velocity as it had at launch. Calculations neglecting air resistance are termed "vacuum ballistics," and they are a very useful tool to understand ballistic performance. In the examples that follow, the ideal performance has been calculated using the vacuum ballistics formulas, and then the actual performance of the system is compared against the ideal flight. These tests were carried out on a deserted beach with one-ounce (437 grain) wooden arrows and a bow with a rated force of 33 pounds at 28 inches of draw. The arrows were actually drawn 30 inches, and launched at an actual velocity of 176 feet per second.

The tests were carried out at launch angles of 15° above horizontal all the way up to a vertical launch. The first tests were carried out at the 15° angle. The ballistic data is summarized in *Figure 1.* While any table has a formidable appearance, the mathematics involved are very simple. The arrow has a velocity *along its flight path* of 176 feet per second. This velocity is broken down into a vertical and horizontal component. The vertical velocity is the velocity along its flight path times the sine of the angle of launch (15°). Thus, V_v = vertical velocity = 176 (.259) = 45.6 feet per second. The horizontal velocity of the arrow is the velocity along the flight path times the cosine of the angle of launch. Thus, V_H = Horizontal Velocity = 176 (cosine 15°) = 176 (.966) = 170 feet per second. Using the vacuum ballistics method of calculation, we find that the arrow will reach its maximum height at the time when gravity has decreased the vertical velocity to zero.

The time to maximum height is calculated by $t = \frac{V_v}{g}$

Flight Shooting at Low Angles

Launch Velocity Feet-Second	*φ Angle of Launch*	*Sine φ*	*Vertical Velocity* $V_v = \frac{Feet}{Second}$	*Cosine φ*	*Horizontal Velocity* V_H	*Time To Maximum Height* $t = \frac{Vv}{g}$	*Maximum Height of Arrow* $H = \frac{Vv^2}{2g}$	*Total Time of Flight*	*Theoretical Horizontal Travel (Feet)*
176	15°	.259	45.6	.966	170	1.42 sec.	32.3	2.84 sec.	483
176	30°	.500	88	.866	152	2.73 sec.	120	5.46 sec.	830

FIGURE 1

The maximum height reached by the arrow can be calculated by either one of two formulas. The easiest way is to square the vertical velocity of the arrow and divide it by two times the acceleration of gravity or 64.4

$$H = \frac{V_v^2}{2g} = \frac{V_v^2}{64.4} = \frac{45.6^2}{64.4} = 32.3 \text{ feet}$$

This gives the maximum height in feet. Another way to calculate the maximum height of the arrow is to square the time to maximum height and multiply it by 16.1.

The total time of flight will be exactly twice the time to maximum height if air resistance is neglected. In the case of the 15° launch, the time to maximum height is 1.42 seconds and the total time of flight 2.84 seconds.

Horizontal Travel of Arrow

The horizontal distance traveled by the arrow is now easily calculated. Neglecting air resistance, there is nothing to slow down the arrow so that it continues its original horizontal velocity of 170 feet per second for the entire time of flight of 2.84 seconds. In the case of the 15° launch, the theoretical horizontal travel is 483 feet. The theoretical and actual performance for the 15° launch is shown graphically in *Figure 2*. The upper graph shows the theoretical range compared to the actual trajectory of a field arrow launched in calm air. The lower graph shows a comparison of the theoretical velocity and the actual velocity of the arrow. The theoretical velocity of the arrow *along its trajectory* is the vector sum of the horizontal and vertical velocities at any point in time. At the beginning of its travel, the arrow is aimed upwards at 15° and has a velocity of 176 feet per second as it is launched. At the extreme height of its trajectory, the vertical velocity is zero; the horizontal velocity is

VELOCITY AND TRAJECTORY AT A 15° LAUNCH

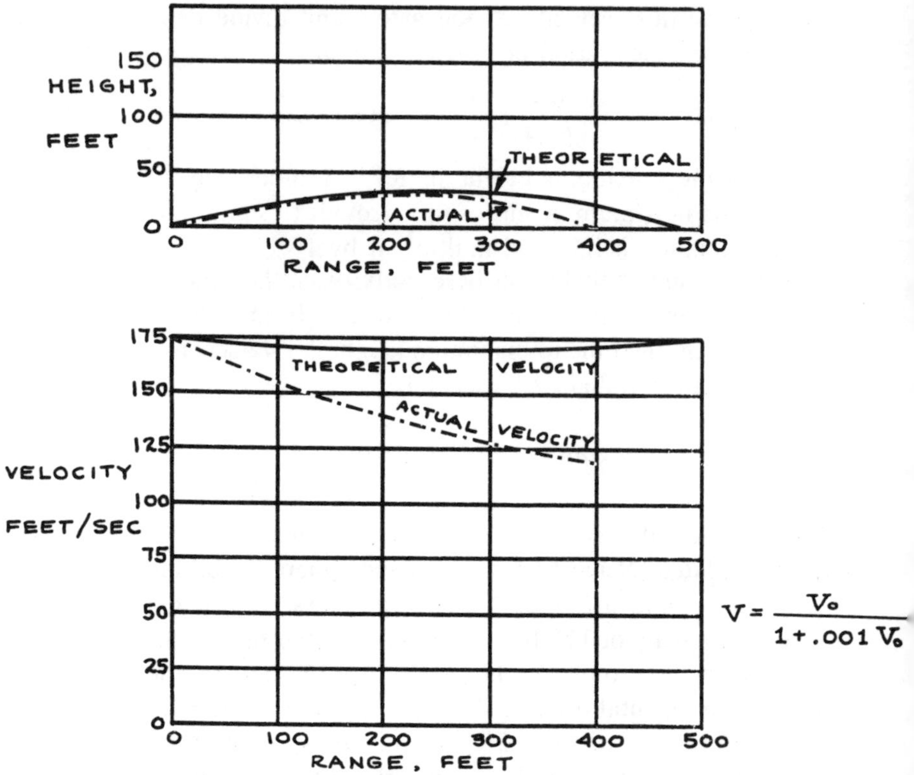

FIGURE 2

170 feet per second, and so the over-all velocity of the arrow at that point when it is traveling only horizontally is 170 feet per second. Using the theoretical conditions of a vacuum, the arrow then accelerates due to gravity and picks up speed again as it comes down the second half of its trajectory. It theoretically hits the ground at 176 feet per second again at exactly a 15° impact angle.

Actual Velocity of the Arrow

The actual velocity of the arrow in flight is quite different. Since an arrow is a rather large projectile and has drag-producing feathers on the back end, the air resistance is quite considerable. The actual flight velocity of wooden one-ounce field arrows fletched with 5-inch feathers is given by:

The formula, $V = \dfrac{V_0}{1 + .001\, V_0 t}$

V = velocity at end of time "t"
V_0 = launch velocity
t = time of flight

This formula will be fairly close for all lightweight wooden arrows. It will not be accurate for fiberglass or aluminum target arrows which have small diameter shafts and small feathers, and a much lower coefficient of drag. The lower curve in *Figure 2* shows that the actual velocity differs greatly from the theoretical velocity due to air resistance. The actual velocity declines steadily and strikes the ground at 120 feet per second. The actual distance traveled by the arrow is 400 feet, or slightly over 80 per cent of the theoretical distance.

ARROW IN HORIZONTAL FLIGHT

An arrow launched horizontally on level ground will travel a very short distance. This is because it will start to drop immediately, and in the first second of flight it will drop 16 feet. This is given by,

$S = \frac{1}{2} gt^2$ where

S = drop in feet
t = time of flight
g = gravity acceleration
g = 32.2

This sad situation exists with rifle bullets or cannon shells, but these high-velocity projectiles will generally cover a third of a mile of range in one second's flight. The arrow, on the other hand, if launched horizontally from a spot 16 feet above the ground level will cover about 160 feet before hitting the ground.

If an arrow were launched horizontally from a high cliff, the only thing that would decrease its *horizontal* velocity would be air resistance. A graph of how this velocity would change is shown in *Figure 3*. The remaining velocity after four seconds is down to about 100 feet per second. The arrow would have traveled horizontally 160 feet the first second, 140 feet the second, 120 feet the third, and 110 feet during the fourth second.

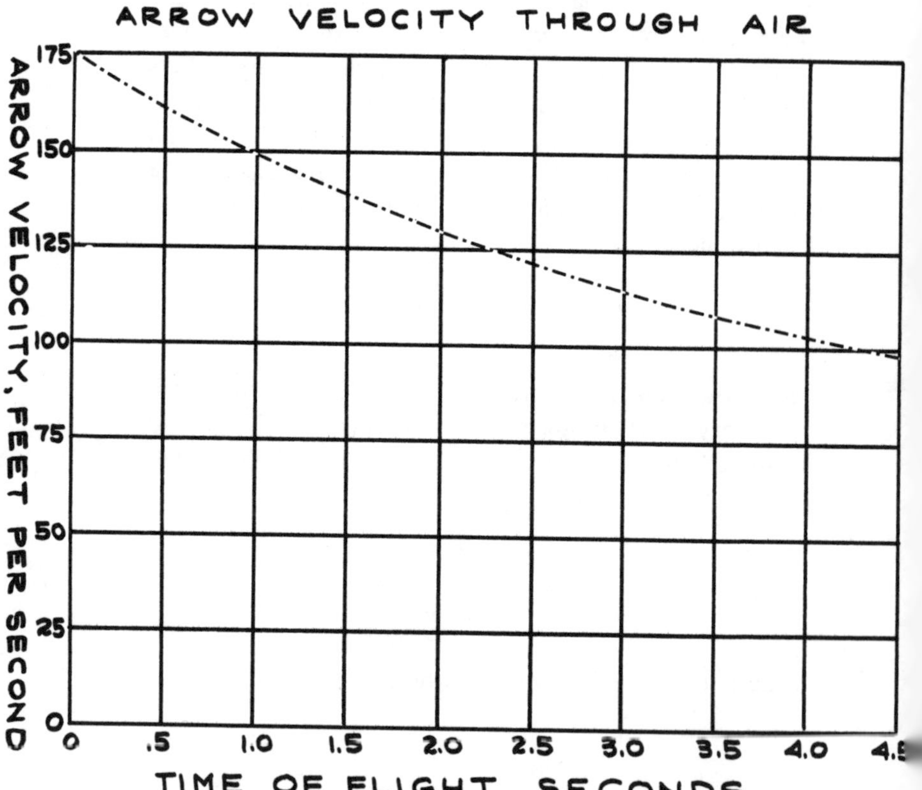

FIGURE 3

ARROWS LAUNCHED AT 30° ANGLE

The next series of tests was carried out on the deserted beach with arrows launched at an angle of 30° above horizontal. These arrows actually traveled an average of 534 feet, or 178 yards. You can follow the flight of arrows over this distance if the light is good, but it is a long distance, and you generally cannot see the arrows after they impact.

A comparison of the theoretical ballistics at 15° launch and a 30° launch is shown in *Figure 1*. With the 30° launch, the vertical velocity is 88 feet per second, and the time to reach maximum height is almost twice as long as with a 15° launch. The arrow rises about four times as high (to 120 feet in the air), and the total time of flight almost doubles, from 2.84 seconds at 15° launch, to 5½ seconds at a 30° launch.

A comparison of the theoretical and actual performance is shown in *Figure 4*. With no air resistance the arrow would travel over 800 feet, reaching its maximum height at the middle of its flight. As the arrow returned to earth, it would pick up speed again and impact at its launch velocity of 176 feet per second. It would theoretically land at an impact angle of 30°. In actual performance, one-ounce wooden field arrows travel an average of 534 feet from the point of launch and impact at an angle of 47°. The comparison of the theoretical and actual curves shows quite clearly that the arrow stays near its theoretical trajectory during the early part of its flight, and then deviates more and more as air resistance slows it down and limits the distance it travels.

The lower graph shows a comparison of the theoretical and actual velocity of the arrow. The theoretical velocity is the velocity along the trajectory of the arrow or the vector sum of V_V and V_H at any point. As the arrow is launched at an angle of 30° to the horizontal, it has a velocity of 176

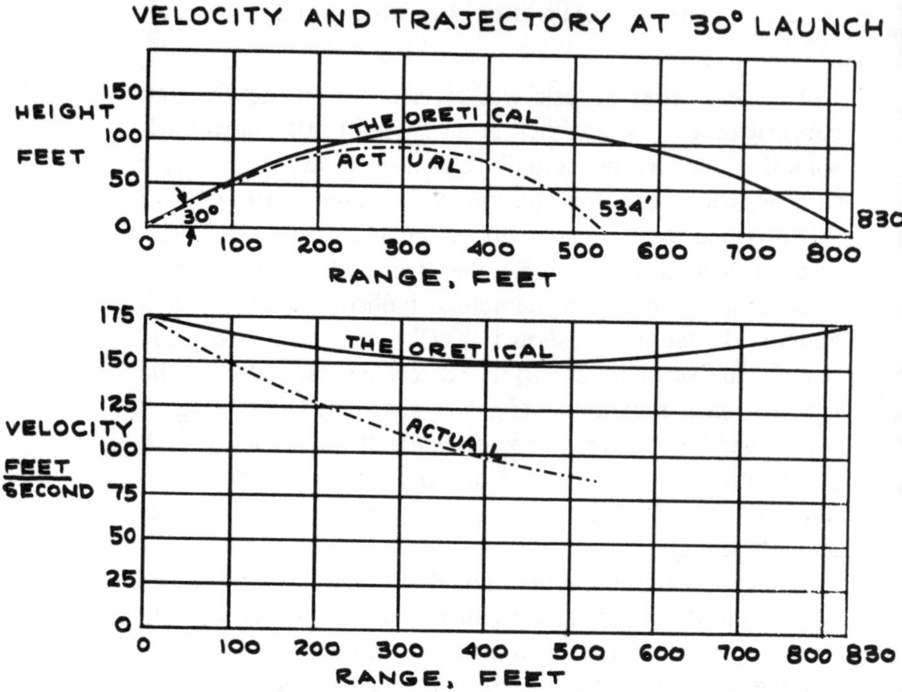

FIGURE 4

feet per second. At the height of its trajectory, the vertical velocity is zero and the arrow is traveling horizontally at a velocity 152 feet per second. As it begins to fall, its theoretical velocity increases due to the acceleration of gravity, and it strikes the ground 830 feet from its point of launch with a velocity of 176 feet per second.

The actual velocity differs radically from the theoretical performance, because wooden field arrows fletched with three feathers have a very considerable air resistance. The actual velocity is plotted in on the lower graph in *Figure 4*. The plot of this velocity differs slightly from the velocity plot for an arrow traveling horizontally. With the arrow shot at a 30° elevation we know, for example, that the velocity decreases more rapidly in the beginning part of the trajectory as the arrow is climbing higher and higher in its trajectory; and, that the velocity would decrease less during the second half of its trajectory where gravity is aiding the arrow's flight.

The arrow launched at 30° has a much longer flight time and travels 134 feet farther than an arrow launched at 15°. The 30° launch is, in fact, a very efficient shot. The arrow travels very near to its maximum range, but with a substantially shorter flight time, and a higher remaining velocity than if shot at 45° launch.

FLIGHT SHOOTING AT HIGH ANGLES OF LAUNCH

The next type of shooting considered is flight shooting at very high angles of launch; that is between 60° to the horizontal and straight up in the air. In this type of shooting, gravity becomes a very strong factor, and we can no longer use one formula to calculate the velocity over the entire flight path. The simplest situation is where the arrow is shot straight up in the air. For obvious reasons, this kind of shooting is not considered safe. It is very easy to lose the arrow against the sky, and even the slightest wind will bring it down anywhere within a 100-foot radius of the firing point. Furthermore, it is very difficult to get a true 90° launch. Normally, you will launch between 80° and 90° and the arrow may come down a good 200 feet away. The simplest ballistic problem of all is the vertical or 90° launch. In this case, the vertical velocity is 176 feet per second, the horizontal velocity is zero. The time to maximum height is about five and one-half seconds, and the theoretical maximum height is 481 feet, as shown in *Figure 5*. The total time

Flight Shooting at High Angles

Launch Velocity	*Angle of Launch*	*Sine* ϕ	*Vertical Velocity* V_v	*Cosine* ϕ	*Horizontal Velocity* V_H	*Time To Maximum Height* $t = \frac{V_v}{g}$	*Maximum Height* $H = \frac{V_v^2}{2g}$	*Total Time of Flight*	*Theoretical Horizontal Travel (Feet)*
176	90	1.0	176	0	0	5.46 sec.	481	10.92	0
176	60°	.866	152.4	.500	88	4.73 sec.	360	9.46	832

FIGURE 5

of flight is 10.92 seconds divided evenly between the travel up and the return to earth. Theoretically, the arrow strikes the earth at 176 feet per second at the exact point from which is was launched. A corollary to these theoretical calculations is that you better not be standing there when it comes down.

A series of measurements, taken with one-ounce field arrows launched at 176 feet per second, gave drastically different results. The arrows reached the peak of their trajectory after an average of only 2.82 seconds. When launched at angles very close to 90° in still air, the arrows remained motionless at the peak of their trajectory for an instant and then fell backwards for 20 or 30 feet and then suddenly flipped over until the point was downwards. The return to earth took an average of 6.5 seconds. The total time of flight was, therefore, an average of 9.33 seconds which was 85 per cent of the theoretical time of flight. What was more surprising is the fact that the arrows spent only 30 per cent of their time going up and 70 per cent of their time of flight returning to earth. If the arrows were launched at an angle of 85° or less, they would reach the top of their trajectory and fall sideways for 10 or 15 feet before rotating into a position where the point was aimed towards the ground. Whether the arrow is falling sideways or falling backwards, the air resistance is obviously far greater than that which would occur during a normal flight.

An analysis of the actual data, secured at launch angles from 60° to 90°, indicates that the second half of the trajectory—the fall of the arrow—may be closely approximated using the formula shown in *Figure 6*. Theoretically, the formula, with no air resistance, would be $V = Vo + 32.2t$, but the air resistance cuts the acceleration to less than half the value in a vacuum or $V = Vo + 14.5t$ for wooden arrows. The distance that the arrow would fall in any period of time after its peak trajectory is shown in *Figure 6.* In order to get an accurate picture of the velocity profile of the falling arrow, it is necessary to break up the time interval into a series of steps. At the end of the first second, the

Velocity Profile of A Falling Arrow

$$V_t = V_o + 14.5t$$

Time interval	V_o	+	14.5t	=	V_t	S =	$7.24t^2$
0	0		0		0		0
1	0		14.5	=	14.5 fs		7.24′
2	14.5	+	14.5	=	29 fs		29′
3	29	+	14.5	=	43.5 fs		65′
4	43.5		14.5	=	58		116
5	58		14.5	=	72.5		181
5.64	72.5	+	9.2	=	81.8		230
6.0	72.5	+	14.5	=	87 fs		260
6.5	87	+	7.2	=	94.2		307
7	87	+	14.5	=	101.5 fs		355′

V_o = velocity at start of time interval "t".
V_t = velocity at end of time interval "t".
S = *Total* distance traveled by end of time interval "t".
t = time since arrow started to fall.

FIGURE 6

arrow is traveling at 14.5 feet per second and has fallen 7½ feet. The velocity gradually builds up during the 6½-second fall to earth, and according to the formula, it has reached a velocity of 94 feet per second and has fallen a total distance of 307 feet. While these figures may not be exact, they check very closely with the actual observed performance of one-ounce wooden field arrows.

Velocity of Arrow Fired Straight Up
(90° Launch) k = .0008

t	$\frac{V_o}{1 + .0008\ V_0 t}$	$32.2t$	V_t	$Vavg$	$\triangle S$
0	176	0	176	—	0
0.5	164.4	16.1	148.3	162.1	81
1.0	154	32.2	121.8	135	67.5
1.5	145.3	48.3	97.0	109.6	42.5
2.0	137	64.4	73	85	30.7
2.5	130	80.5	49	61	55
2.82	126	90.8	35	42	13.5
					290.2′

V = Remaining velocity after time "t."
t = time in seconds after launch.
Vo = launching velocity.
k = .0008 for 1 ounce wood field arrows.
Vt = velocity after time "t".
Vavg = Average velocity over time interval (1/2 sec.).
ΔS = distance traveled by arrow over time interval (1/2 sec.).

$$V = \frac{Vo}{1 + .0008\ Vot.} - 32.2t$$

FIGURE 7

Rise of the Arrow

In trying to understand the actual performance of an arrow during its upward path, we must take into account both the air resistance and the deceleration due to gravity. The formula to calculate the actual performance of one-ounce field arrows is shown in *Figure 7*. In this case, the deceleration factor is slightly less than for horizontal travel. In order to fully understand the performance, it is necessary to break up the calculations into time intervals of about a half-second each. The time is shown in the first column. The velocity change due to air resistance is shown in the second column. The loss in velocity due to the deceleration of gravity is shown in the third column, and the velocity at the end of the time interval taking both of these factors into account is shown in the fourth column. In order to calculate the distance traveled, the average velocity over the time interval is calculated by averaging the velocity at the beginning and at the end of the time interval. When the averaged velocity is multiplied by the length of the time interval (half of a second) the distance traveled by the arrow in that time interval is secured. The calculations for a 90° launch show that the arrow has theoretically a velocity of 35 feet per second remaining at the end of the time interval, and has reached a height of 290 feet in the air. The formula is slightly in error since the velocity should be zero, but it is a relatively close approximation of the observed performance.

By comparing *Figures 6* and *7,* the velocity changes in an arrow during its upward flight and fall back to earth may be visualized. Instead of attaining a theoretical height of 480 feet, the arrow actually reaches about 300 feet. Instead of equal time up and down, less than three seconds is used in the upwards flight, and 6½ seconds on the return to earth.

Arrows Launched at 60°

If an arrow is launched at an angle of 60°, then gravity remains the major factor slowing the arrow down in its flight. For any angle between 60° and straight up, this formula may be used to calculate the velocity of the arrow up to the peak point of its flight.

$$V = \frac{V_o}{1 + .0008\,V_o t} - 32.2t$$

V = remaining velocity
t = time of flight
V_o = vertical velocity at launch

These calculations worked out for one-ounce field arrows launched at a velocity of 176 feet per second are shown in *Figure 8*. The vertical velocity of the arrow at the time of launch is 152 feet per second. At the end of one second, this has decreased to only 103 feet per second, and at the end of two seconds it is down below 60 feet per second. There is a slight inaccuracy in the formula at a launch angle as low as 60°, since the calculations show that a velocity of nearly 40 feet per second remains at the high point of the trajectory, where in fact the velocity is zero. The height to which the arrow travels checks very well with the graphical results shown in *Figure 9*.

Vertical Velocity of Arrow Fired at 60° Angle

$$V_t = \frac{V_o}{1 + ktV_o} - gt$$

t	$\frac{V_o}{1+ktV_o}$	−	gt	=	V_t	Vavg	△ S
0	152.4		0		152.4	—	0
.5	143		16.1		127	139.5	70′
1	135.4		32.2		103	115	57.5
1.5	128.8		48.3		80.5	91.7	46.
2.0	122.5		64.4		58.1	69.3	34.6
2.44	117.4		78.6		38.8	48.4	21.
						Total travel =	229 ft.

V_t = Vertical velocity at end of time interval "t".
g = Deceleration due to gravity.
Vavg = Average velocity of ½ sec. time interval.
△ S = Distance traveled over ½ sec.
k = .0008 for wooden one ounce field arrows.

FIGURE 8

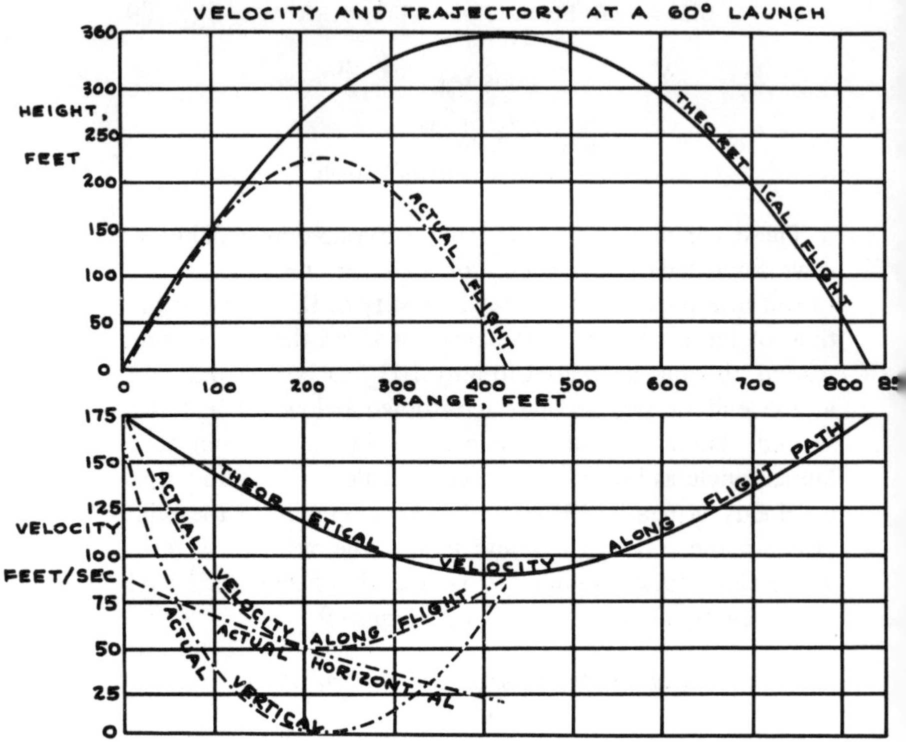

FIGURE 9

Fall of the Arrow

The second half of the trajectory is much like the falling of an arrow fired vertically. The arrow reaches the peak of its trajectory and seems to hang there. Then it starts to fall sideways and gradually noses over into a point down flight path. The fall of the arrow may be approximated by the formula shown in *Figure 6*. In the case of a 60° launch, the arrow falls for 5.64 seconds, and attains a velocity of about 82 feet per second at time of impact. The buildup in *vertical* velocity is almost identical to the arrow falling from a vertical launch.

The theoretical performance of an arrow launched at 60° is shown in *Figure 5*. The vertical velocity is much greater than the horizontal velocity, but the theoretical performance is impressive. The arrow describes a gigantic arc reaching a height of 360 feet in the air, and traveling 832 feet horizontally before impacting the ground.

The comparison of theoretical and actual performance is shown in *Figure 9*. The contrast is dramatic. The arrow reaches only half of its theoretical horizontal travel, and reaches 230 feet in height, compared to a theoretical height of over 350 feet.

Velocity of Arrow Along Flight Path

The actual velocity of the arrow along the flight path is the vector sum of its actual vertical velocity and actual horizontal velocity. A comparison of the theoretical velocity along the flight path and the actual velocity along the flight path is shown in the lower charts of *Figure 9*. The actual vertical velocity starts at 152 feet per second and decreases to zero at the height of the trajectory. During the second half of the trajectory, the arrow builds up a vertical velocity from zero to 81.8 feet per second at the end of 5.64 seconds. But the arrow also has a horizontal velocity during its entire flight path. This horizontal velocity averages 52 feet per second over the flight distance. The actual horizontal velocity can be calculated from the formula shown in *Figure 3*, using a Vo of 88 feet per second. The actual horizontal velocity is shown graphically starting at 88 feet per second, and decreasing due to air resistance to a final velocity of slightly under 25 feet per second. The actual velocity of the arrow along its flight path will be the vector sum of the horizontal and vertical velocities. The vector sum for any point in the flight path may be found by calculating the vertical velocity, squaring it, and adding the squared horizontal velocity. The velocity along the flight path will be the square root of these two sums:

$$V = \sqrt{V_H^2 + V_V^2}$$

It would be ideal if there was a very simple way to look at the velocity picture and such simple systems can be used at angles up to 30° and at angles over 80°. For angles between 30° and 80°, neglecting either the horizontal or vertical velocity leads to very serious errors; and this more complex technique must be used to understand what the arrow is really doing in flight.

ARROW LAUNCHED AT 45° ANGLE

The maximum range with any projectile is theoretically attained with a 45° launch. Specialized flight-shooting equipment has been used to attain extremely long range with 45° launch of arrows. Ranges of 600 yards have been attained with bows rated at 50- pound draw, *but* these ranges are attained with small diameter arrows with minute feathers, and bows with short limbs which are not handled in a normal fashion.

The arrows used in these flight-shooting tests were normal field arrows, 3/8 inch in diameter with three feathers each 5 inches long. The combination of large diameter and large feather makes a rather high drag. The effect is a serious reduction in actual range compared with theoretical range.

The theoretical height attained by the arrow is 240 feet, the actual about 155 feet. The theoretical distance to impact is 962 feet, the actual only about 505 feet average, or 52 per cent of the theoretical distance.

Actually, 45° does not give the maximum range with normal archery equipment. The actual distances for the angles near 45° were as follows:

Angle	*Range to Impact, feet*
40°	532
45°	505
50°	483

The optimum angle to achieve maximum range is about 42° with normal equipment.

The performance of the one-ounce wood arrows at a launch velocity of 175 feet per second is shown in *Figure 10*. The theoretical range is shown in the upper graph, and the actual flight is plotted immediately below it. The high drag of these arrows shortens the flight tremendously.

In the lower graph of *Figure 10*, the velocity characteristics of the system are shown. The theoretical velocity *along the flight path* starts at 175 feet per second, then drops to the horizontal velocity (V_H = 125 ft/sec.) at the peak of the trajectory where the vertical velocity is zero. The velocity then builds back up on the second half of the trajectory until the arrow hits the ground at 175 feet per second at a 45° angle.

The actual velocity relationships are very different. While there is not enough data to perform an exact analysis, a close approximation can be made. The vertical velocity of the arrow at launch is:

$$V_v = 175 \times \text{sine } 45° = 125 \text{ ft/sec.}$$

At the peak of the trajectory, the arrow is slightly over 150 feet high in the air. The fall of the arrow to earth during the second half of the trajectory can be closely approximated by:

$$S = \text{Distance of fall} = 155 \text{ feet} = 7\tfrac{1}{4}t^2$$

From this, the time of fall can be calculated as $t = 4.7$ seconds. The velocity can be approximated by:

$$V_v = V_o + 14.5t$$

The vertical velocity is zero at the height of the trajectory so the impact velocity becomes:

$$V_v = 0 + 14.5\ (4.7)$$
$$V_v = 70 \text{ feet/second}$$

The horizontal velocity of the arrow starts out at:

$$V_H = V_o \text{ cosine } 45° = 175 \times .707 = 125 \text{ ft/sec.}$$

The *only* force slowing down the horizontal velocity is air resistance, and this is given by:

$$V = \frac{V_o}{1 + .001\ V_o t} = \frac{125}{1 + .125t}$$

The best way to determine the time of flight is to have an observer with a stopwatch. An average of five or ten shots is a very accurate measure. An observer standing well to one side can also measure the time to maximum height, and the time from maximum height to impact with the ground. With this data, you can see the performance of your bow and arrow system quite clearly.

The calculations applied to this flight are shown in the lowest graph shown in *Figure 10*. The actual velocity decreases very rapidly during the first part of the trajectory, and then remains more nearly constant during the arrow's return to earth.

AIR RESISTANCE OF ARROWS

The flight of projectiles through the air has been a subject of study for hundreds of years. As long as muzzle-

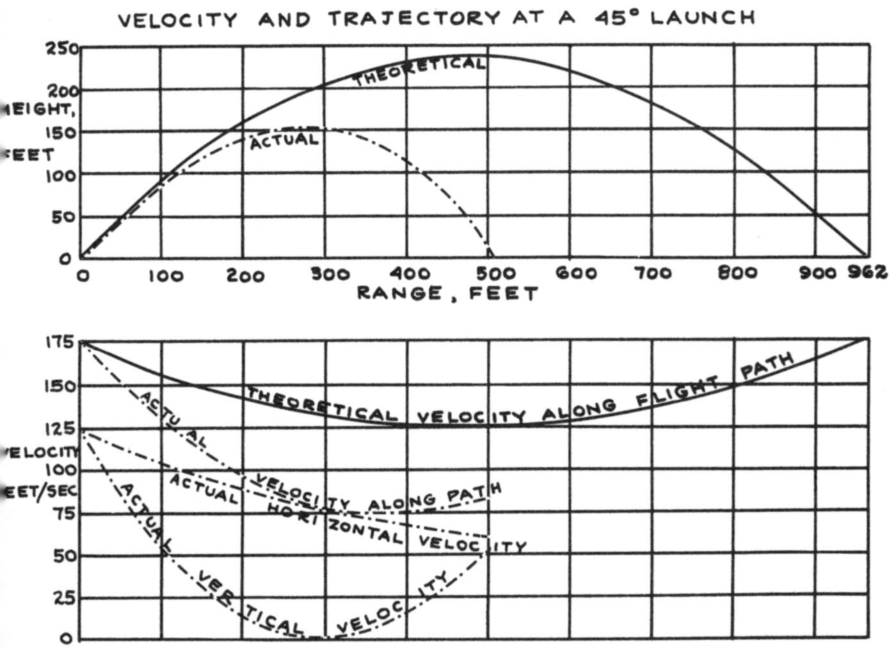

FIGURE 10

loading, smoothbore cannons were used, the flight of a projectile was difficult to predict at long range. The cannon ball literally bounced from one side of the cannon bore to the other, and its direction of flight was influenced by the last bounce before leaving the muzzle.

As late as the American Civil War, smoothbore cannons were in wide use, and they gave excellent service at ranges up to a mile. Rifled cannon were also in use in the Civil War. They required a longer time to load, but were more accurate at all ranges. During the second half of the nineteenth century, the science of ballistics received increasing attention and government-sponsored test programs were carried out in many countries. Scientific textbooks were written based on these tests, and exterior ballistics became a well-developed science so that by 1917 artillery duels with sixteen-inch naval cannon could be carried out at ranges of up to fifteen miles.

In the battle for Leyte Gulf, in World War II, United States ships opened fire at a range of 15,600 *yards* and achieved hits with the first shots fired. Other engagements during this battle were fired at ranges up to 26,000 yards (thirteen miles) with effective results. This is incredibly accurate shooting by any standards.

One of the factors receiving close attention over the past century is the air resistance of projectiles. The actual method used is to carry out extensive firing experiments, and then to construct elaborate tables from which the performance of a wide range of projectiles can be predicted.

The factors affecting the drag of a projectile at low velocity are:

$$\text{Drag} = P\,A\,V^2\,k$$

P = Air Density
A = Area of largest cross section of the projectile
V = Velocity
k = Form factor

Examination of this formula shows some of the problems in calculating air resistance of arrows accurately. Some dimensions and weights of nine different arrows are shown in *Figure 11*. The overall length and weight of different styles of arrows differs widely. The rated length of the arrow is always shorter than the actual length.

The overall length of target arrows is measured from the bottom of the neck to the end of the cylindrical section of the point. Thus the target arrows of any material will be the shortest style.

Aluminum arrows, shown as the first three entries on

TYPICAL ARROW DIMENSIONS

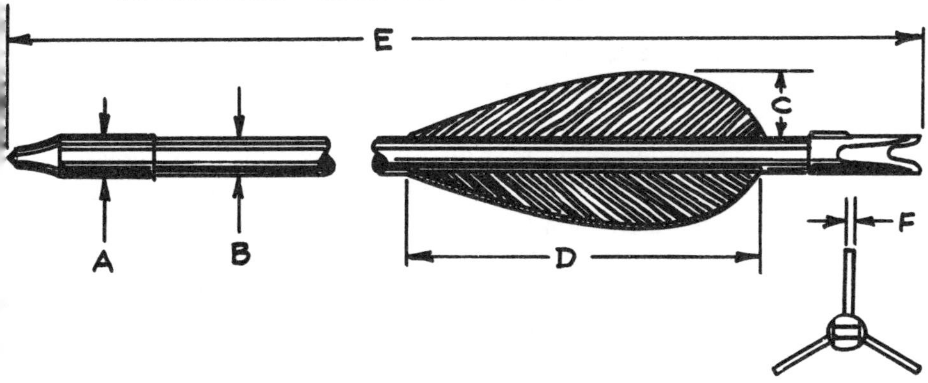

	A	B	C	D	E	F	WEIGHT
1. ALUMINUM, TARGET 29" LONG	.267	.267	.500	3.75	29.75	.050	413 GRAINS .94 OZ
2. ALUMINUM, FIELD 29" LONG	.325	.315	.550	4.87	31.30	.045	590 GRAINS 1.35 OZ
3. ALUMINUM, HUNTING 29" LONG	1.10	.315	.550	4.87	32.75	.045	663 GRAINS 1.52 OZ
4. FIBERGLAS, TARGET 29" LONG	.301	.296	.580	4.00	29.50	.050	392 GRAINS .90 OZ
5. FIBERGLAS, HUNTING 28" LONG	1.10	.322	.550	4.85	31.62	.040	640 GRAINS 1.46 OZ
6. FIBERGLAS, FIELD 28" LONG	.350	.322	.550	4.85	30.30	.040	571 GRAINS 1.30 OZ
7. WOOD, TARGET 28" LONG	.320	.314	.580	3.25	28.75	.040	300 GRAINS .68 OZ
8. WOOD, FIELD 29" LONG	.374	.345	.500	5.00	31.44	.055	437 GRAINS 1.00 OZ
9. WOOD, HUNTING 29" LONG	1.10	.355	.520	5.00	32.80	.050	539 GRAINS 1.23 OZ

FIGURE 11

the list, are the heaviest arrows of each class. The target arrows weigh nearly an ounce, even though the shafts are only .267 inches in diameter. The overall length is only three-quarters of an inch over the "rated" length of the arrow. The combination of small diameter and high weight gives the aluminum target arrow great range and penetration.

The aluminum hunting arrow and field arrow are considerably longer and heavier. They are built with a 5/16 inch diameter shaft, and the hunting arrow with a rated length of 29 inches has an actual overall length of 32.75 inches. This discrepancy in length seems strange until you think about it. A target arrow may be safely drawn until the point is on the arrow rest, or on your left hand if you are a bare-bow enthusiast. A hunting arrow is fitted with razor sharp edges on the point, and the extra length is to ensure that these edges are safely away from your hand.

The aluminum hunting arrows are the heaviest of the arrows measured. With a four-bladed broadhead the arrows averaged a full 1½ ounces, or 663 grains. This increased weight requires a heavy bow to provide adequate velocity levels. The added weight also provides a much more rugged arrow to stand the abuse of the hunting field.

The story is similar for the wood and fiberglass designs. The target arrows are the shortest and lightest design. The fiberglass target arrows are close to the aluminum arrow in weight, but have a larger shaft at .296 inches diameter. The wood target arrows are extremely light, since these arrows are spined for a 25-pound bow. The 300-grain arrows will give good velocity and accuracy from a light bow, even at ranges of 50 yards.

The wood field and hunting arrows are much more massive. The shaft diameter is close to ⅜ inch, and the weight is one ounce to 1¼ ounce. The field and hunting arrows are designed for much heavier bows, and they must be strong enough to stand the high initial acceleration from the bow as well as the collision with an animal or tree at the other end of the flight. The wood field arrows listed as entry number eight in *Figure 11* are the arrows used in the flight-shooting tests in this chapter. The combination of larger shaft diameter and large feathers gives these arrows a relatively high drag compared to other arrows on the chart.

In conducting long series of tests to gather data mishaps have occurred with many different types of arrows. The most disastrous events were when arrows in full flight struck a stone wall which borders the author's property in Connecticut. The arrows reacted in different ways, depending on the material used for the shaft. Aluminum arrows were always totally ruined, for the point was mashed, deforming the front of the shaft, and the shaft was bent too badly to straighten. Wood arrows invariably broke off immediately behind the point. Some had split shafts, and these were totally ruined. The wood arrows which broke behind the point were repaired by cutting off the damaged section, sanding a new taper to the wood, and glueing on a new point. These repaired arrows were shorter than the remainder of the set, but can be used for field-shooting practice where loss and damage are frequent.

Fiberglass arrows proved to be very tough. On direct impact with a hard surface at high velocity the point of the arrow was driven up into the shaft. Technically the fiberglass fails in tension around the periphery of the shaft, and the end looks like a peeled banana. Even when a terrific impact occurs the failure extends back for only about an inch, and this may be repaired by cutting off the damaged portion and glueing in a new point. In general the fiberglass arrows seem to be able to absorb at least twice as much punishment as either wood or aluminum arrows before suffering permanent damage.

Any arrow that glances off the target, or a rock, should be examined very carefully before further use. Check to be sure that the cement bond between the feather and shaft has not been broken. If the leading edge of the feather has started to separate from the shaft, the feather is likely to be ripped off if shooting is continued. Wood arrows should be checked for a fractured shaft immediately behind the point. Both wood and aluminum arrows should also be checked for straightness after a glancing blow. Wood arrows can be straightened by holding the bent area in the steam jet from a kettle, and then bending the shaft in the opposite direction with the fingers. The steam from a kettle is also very handy for restoring battered feathers. The feathers can be restored by twirling them rapidly in the steam jet for just a few seconds.

The air resistance of an arrow may be estimated by referring to the data shown in *Figure 11*. The cross section area of the arrow is the sum of the area of the shaft plus the area of the three feathers. The frictional area of an arrow is much greater than a bullet due to the long shaft, and the relatively large area of the feathers.

The drag acts to slow down the arrow. The rate at which it slows down the arrow is the *deceleration*, and this may be found from:

$$a = \frac{D\,g}{W}$$

a = deceleration
D = Drag
g = gravity
W = Weight of arrow in pounds

By combining this formula and the previous formula for drag:

$$\text{Drag} = P\ A\ V^2\ k$$

we have: $$\text{Deceleration} = a = \frac{K\,P\,A\,V^2\,g}{W}$$

This formula tells us that the heavier the arrow, the better it will retain its velocity. The formula may be further refined to show the velocity change over a period of time "t."

$$V = V_o - \frac{K\,P\,A\,V_a^2\,g\,t}{W}$$

The difficulty in using this formula is that you are working with three different velocity figures.

V = velocity at end of time "t"
V_o = velocity at start of time interval
V_a = average velocity over time interval
K = drag factor
P = air density in pounds/ft-inch^2
(This is an odd unit to make the dimensions come out right. At 70°F at sea level P = .00052)
A = largest cross section area of arrow
W = weight of arrow in pounds
t = time interval, seconds
g = gravity = 32.2 ft/sec^2

The drag equation applied to a one-ounce wood field arrow launched horizontally from a high cliff, and considering only the horizontal motion, is shown below. The calculations were based on ½ second time intervals. Note that ΔV is the velocity loss over the time interval, and that V, the velocity at the end of the time interval, becomes Vo for the next time interval.

The horizontal distance traveled by the arrow over the ½ second interval is given as ΔS, and the cumulative horizontal distance from launching is given as S. The calculations are tabulated in *Figure 12*.

This data is shown graphically in *Figure 3*. The rapid decline in drag as the velocity decreases is clearly shown in

Time After Launch (Seconds)	*Initial Velocity Vo*	*Average Velocity Va*	*Drag, Pounds D*	*Velocity Loss* ΔV	*Velocity At End of Time Interval V*	*Distance Traveled in Time Interval* ΔS	*Total Distance (Feet) S*
.5	176	169	.055	14.3	162	84	84
1.0	162	155	.048	12.3	149	78	162
1.5	149	143	.041	10.6	139	72	234
2.0	139	134	.035	9.2	130	67	301
2.5	130	126	.031	8.0	122	63	364
3.0	122	118	.027	7.1	115	59	423
3.5	115	111	.024	6.3	108	56	478
4.0	108	105	.022	5.7	105	53	531

FIGURE 12

this table. In looking at these formulas, and the example shown in *Figure 12,* the factors which would increase range become apparent:

1. Decreasing cross-section area of arrow will reduce drag. The specialized arrows used in flight shooting are very small in diameter, and made of compressed cedar or other very strong materials. The feathers are minute, to minimize drag. The cross section area of the wooden field arrows shown as type 8 in *Figure 11* is:

$$\text{Shaft area} = \frac{\Pi d^2}{4} = \frac{3.14}{4} \times (.376)^2 = .111 \text{ in}^2$$

$$\text{Three feathers} = 3 \text{ x } .055 \text{ x. } 500 \qquad = \underline{.083 \text{ in}^2}$$

$$\text{Total area} \qquad .194 \text{ in}^2$$

   The cross-section area of the aluminum arrows (Type 1), by comparison is only .131 in^2 or 32 per cent less. Thus, we would expect the aluminum arrows to carry much farther than the wood, and, in fact, they do have much greater range.
2. Decreasing the air density will decrease drag. Most of the new flight shooting records are established at ranges located at high altitudes, where the air offers less resistance.
3. Improving the shape of the projectile to reduce the drag factor K will increase the flight.
4. Increasing the weight of the projectile, without decreasing the velocity, will increase range.
5. Increasing the velocity of the arrow, without decreasing weight, will increase range.

While long-range, accurately instrumented firing data is not available to define deceleration factors for a wide range of arrows, the drag formulas do show the relationships that affect arrow ballistics. Flight shooting with a bow and arrow can be an interesting afternoon's sport. The arrows used in these examples have very high drag, so that care should be used in early firing tests to determine the range of the system. Aluminum and fiberglass arrows will tend to cover a greater percentage of their theoretical range than the wooden field arrows used in these examples.

Figure 13 shows a young archer enjoying himself at flight shooting. It is great summer sport on an empty beach to watch the arrows sail gracefully through the air and strike an impact point down range. The young archer may not be too interested in the exact range to the nearest foot or the exact angle of impact of the arrow, but he does enjoy seeing how far the arrows will go at various angles of launch.

Young archers can enjoy distance shooting.

FIGURE 13

A view down the flight-shooting range is shown in *Figure 14*. Old iron posts have been set in the sand at 50-foot intervals at both ends of the range. The center portion of the range has been left blank since no arrows impact in this area. A 100-foot steel tape was used to measure off the distances from the launching point. Any identifying mark in the sand can be used, but it is well to clearly identify the range from the firing point. On this range, the distance was written on slips of paper and jammed into the spaces in the old iron posts to insure that the distances were accurately recorded.

A view down the 250-yard range. Markers have been driven into the sand every 50 feet. The angle of launch is measured by an observer to the left of the archer.

FIGURE 14

The author is shown letting an arrow fly down range with a modern laminated bow. At medium angles of launch, the arrow often became invisible at the top of the trajectory. The arrow became briefly visible again during the fall and then was invisible again after impact in the sand. It is a wise precaution to walk over the course just before doing flight shooting to insure that there is absolutely no chance of unwary spectators anywhere near the impact area. Generally, the visibility on a sandy beach is excellent for great distances in all directions and this is a very safe and enjoyable sport.

If you wish to carry out tests to establish distances for your particular archery equipment, it is well to establish a regular procedure. An excellent way is to plan to do your flight shooting in the early morning or late afternoon around five o'clock when the wind tends to be at a minimum. An observer should stand to one side with an angle-measuring device and tell the archer to move his bow up or down until the proper launch angle is achieved. If the observer has a stop watch to time the flight, this provides additional information. Launch at least ten arrows under exactly the same conditions. This will give a good statistical average distance. After firing, walk out to the impact point with the 50-foot tape. Measure the distances to the arrows from the nearest reference line. It is a good idea to identify each arrow with a painted number, and then record both the number of the arrow and the distance of flight. Then stand at the nearest reference line and shoot the arrows back towards the original launching point. This will average out the effect of any slight wind that is blowing. If a heavy wind is blowing, the results will be meaningless since the wind has a very serious effect on the long flight of the arrows.

4
Arrows and Accuracy

LOW VELOCITY PROJECTILES

There is evidence that the bow and arrow as a system for launching projectiles goes back at least 20,000 years. The first arrows probably consisted of a relatively straight stick sharpened at one end by rubbing on a rock and with a simple notch at the other end. The bow was a simple wooden bow with animal sinew for a string. There were no feathers on the arrow, and the range of the system was relatively short.

The ancient bow reached its greatest technical development in the Far East where reflexed bows with a composite structure of wood and animal sinew were developed and widely used. These bows were short in length, but were constructed in a manner similar to the recurved modern hunting bows.

The bow and arrow system in Western culture reached its greatest use in the 700 years following the Battle of Hastings in 1066. At this battle, the Normans under William the Conqueror found they could not break the English

Army on the hill at Hastings. Late in the day, William decided to use his archers to provide a mass fire by shooting the arrows high in the air and dropping them behind the wall formed by the Englishmen-at-arms with their shields. The high-angle fire was effective, fatally wounding King Harold of England, and deciding the outcome of the battle.

In the period following this battle, the British passed archery laws specifying that every able-bodied male between sixteen and sixty must have weekly target practice with the bow and arrow, excluding the nobility. The standard English bows were 6 feet 4 inches in length, and the law specified that they must be manufactured by the various towns and made available for private purchase at the price of one shilling.

The arrows used with these English bows were referred to as "cloth yard shafts," and were from 27 to 36 inches in length. The law also specified that the system must have sufficient power to drive a 36-inch arrow through a 4-inch oak door at 100 yards so that the shaft protruded on the far side "a hand's breadth." These bows required constant training to develop and maintain the muscles required for these power levels. Although great feats of accuracy have been reported historically, particularly in the legends of Robin Hood and his men, the effectiveness of the bow-and-arrow system was not dependent on individual accuracy. Volley firing was used to send a cloud of arrows towards the enemy. The simultaneous launch of a large number of arrows which had great power and range was a devastating military weapon.

In America, the bow and arrow had an equally interesting history. The bow and arrow and lance were the two projectile systems used by the Indians, and of these the bow and arrow was a far more effective system. While the stories of the accuracy of Indian shooting have not reached the legendary proportions of Robin Hood and his men, there has been considerable distortion of the facts. The Indians were relatively poor shots with their bow and arrow, and an examination of ancient Indian arrows shows clearly why this is so. In hunting, the Indian used cunning and stealth to approach his game within point blank range before releasing his arrow. Mounted Indians hunting buffalo on the western plains generally rode into a range of between 4 and 10 feet before firing their arrow at the running buffalo.

The most feared Indian warriors in the nation were the Apaches who terrorized the Southwest from the coming of the white man until the early 1880's. The Apaches also developed the most sophisticated archery equipment. Examples of footed Apache arrows in the Arizona State Museum in Tucson show great care and technical skill. The construction of these footed arrows is shown in *Figure 1*. The beautiful examples shown in the Arizona Museum have, of course, been selected from among thousands of arrows and represent the very finest examples remaining. A typical arrow consists of a chipped flint point which was tightly attached to a hardwood foot. One technique was to make the connection with damp rawhide which then tightened and made a very strong joint when the rawhide dried out. Because there was a desperate shortage of satisfactory wood in the Southwest, the hardwood foot on the arrow is relatively short and the main shaft is constructed of cane. The hardwood was fitted inside a hollow section of the cane which has been split for a short length and then the joint wrapped with rawhide or sinew. The feathers were also held on with a rawhide or sinew binding. The Apache arrows were generally about 30 inches long, and used mountain mahogany for the hardwood foot. Where wood supplies were more plentiful, the Apaches also made their arrows with all hardwood shafts. An examination of the arrow shown in *Figure 1* would reveal very quickly one of the limitations on accuracy. As the knobby cane surface bumped its way past the bow, certain inaccuracies in launch would almost certainly occur.

The highest development in wooden arrows during the past century used much the same techniques as the Apache arrow. Footed wooden arrows were produced in which a hardwood front section was spliced to a cedar shaft, and then a hardwood tail was added on the arrow into which the notch was cut. As late as 1950, many people considered these the finest arrows available, and used sets of footed wooden arrows are still occasionally available. Footed arrows are necessarily much more expensive to produce than standard wooden arrows, and they must be very carefully constructed to make a shaft that is strong and true. The footed wooden arrow has been almost entirely displaced by fiberglass and aluminum arrows.

Modern Wooden Arrows

The wooden arrow remains the standard arrow for archery in the United States, and the vast majority of arrows sold are of cedar construction. A typical wood target arrow is shown in *Figure 2*. The arrow shown in the illustration is fine for a light bow. A 5/16-inch cedar shaft is used, and the maximum diameter of the arrow is over the target point. The target point is either a turned steel part made on a screw machine, or else is a deep drawn cup made on a machine similar to those used for making bullet jackets. The point is retained on the arrow by either knurling or glue or both. The back of the shaft is tapered to a conical point, and a plastic nock is glued to the wooden shaft.

This construction has many advantages over the old

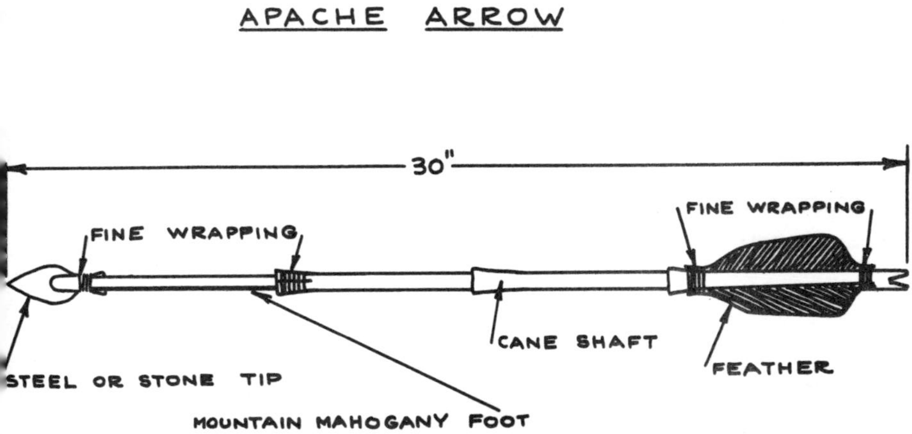

FIGURE 1

method where the nock was cut directly into the wood. The plastic nocks are very tough. They are a complex shape with flats on the sides and a ridge on the top which allows the experienced archer to properly orient the arrow just by feel. If the nock does break, the arrow is not ruined, and a new nock may easily be installed. Unbelievable as it seems to the beginner, modern archers can shoot so accurately that one arrow will occasionally hit the nock of another arrow already in the target and break the nock.

Hundreds of thousands of wooden target arrows are made each year from carefully selected Port Orford cedar. The very finest shafts are used in matched target arrows, which are sold in boxed sets of a dozen. All of the arrows

CONSTRUCTION OF WOODEN ARROWS

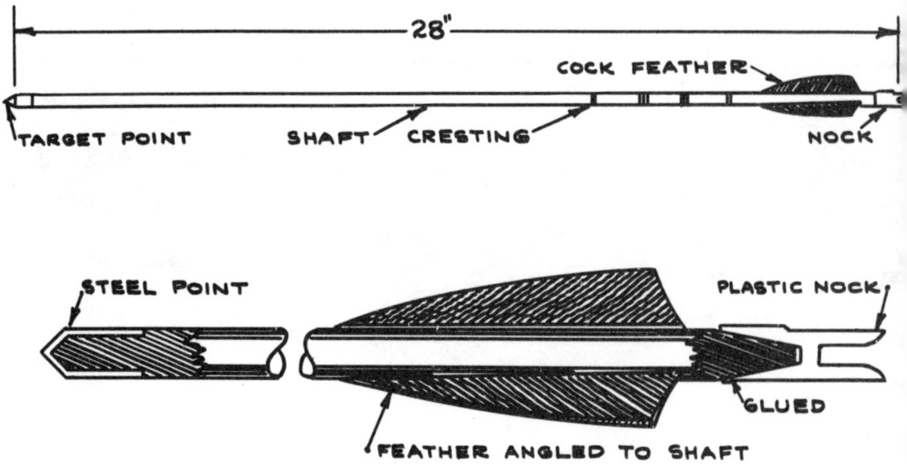

FIGURE 2

within a set match each other very closely in straightness, stiffness and weight. The accuracy of these arrows is usually excellent.

The arrows which do not meet these very high standards are generally sold as economy sets, while the bulk arrows sold from open racks in the archery and department stores have the lowest standards of straightness.

Target arrows are guided by three turkey feathers glued to the shaft. The feathers are generally about three inches long, slightly over half an inch high, and have a thickness of between .035 and .045 inch. They are very carefully glued to the shaft in a special jig, which gives them a slight angle to the axis of the shaft. This twist causes the shaft to revolve slowly on its way to the target. The twist on a target arrow is generally very slight, and the small feathers provide a low drag.

The rated length of a target arrow is the distance between the groove in the nock and the end of the parallel section on the point. The over-all length of the arrow is about ¾ inch longer. The weight of a 28-inch target arrow spined for a light bow is only about two-thirds of an ounce, or 300 grains. Since the drag of an arrow is proportional to the cross-sectional area, these target arrows with their small dimensions have low air resistance. The velocity of the target arrows will be quite high. One-ounce field arrows, launched from a 40-pound bow draw 28 inches, will have a launch velocity of 176 feet per second. This same energy applied to the light target arrows will give an approximate launch velocity of

$$V = 580\sqrt{\frac{F}{W}}$$

F = rated draw of bow.
W = weight of arrow in grains.
(7,000 grains = 1 pound)

$$V = 580\sqrt{\frac{40}{300}}$$

$$V = \frac{580}{2.74} = 212 \text{ feet/second}$$

This relatively high velocity has a further advantage in reducing the drop and time of flight. Thus, at a 50-yard range, the one-ounce field arrow has a flight time of .95 seconds, and a drop of nearly 15 feet. The lighter target arrow has a time of flight of ¾ second, and a drop of only 10 feet.

A wooden field arrow differs from the target arrow in having a much sturdier shaft. The shaft diameter is generally about .350 inch, and the diameter over the point is a full ⅜ inch. Since the point is heavier and the shaft larger, the feathers are also usually about 5 inches long with a height of about ½ inch.

The construction of a typical aluminum or fiberglass arrow is shown in *Figure 3*. The illustration has been drawn by the author from sample aluminum and fiberglass arrows supplied by Herter's Incorporated. A cross section through a hunting arrow is shown. The aluminum and fiberglass shafts are hollow, and the wall on some of the aluminum shafts is only .015 inch thick (1/64 inch thick). At the front of the shaft an aluminum insert is cemented in. The insert has a cylindrical section which is glued into the fiberglass shaft, or pressed into the aluminum shaft. The front of the insert is machined to form a 5-degree taper. Any type of arrow point may be cemented to this tapered section. A hunting broadhead is shown in this illustration. The heart of the Herter style broadhead is a tapered cup which has been machined from solid steel. Two hard, steel blades have been copper brazed to the cup at 2000 degrees in a furnace. The blades should be sharpened to a razor's edge before use in the hunting field. Field points or target points may also be cemented on the taper.

In order to balance the arrow with a heavy hunting broadhead, five-inch-long feathers are used for fletching. The large feathers do increase the air resistance, so that the arrow does not retain its velocity as well as a target arrow of equal weight. The large feathers also make the arrow more sensitive to a crosswind, but they are absolutely essential to maintain stable flight with a heavy point.

The back of an arrow is constructed much the same as the front. An insert with one cylindrical end and one tapered end is used. The cylindrical end is glued or driven into the hollow shaft. A plastic nock is then glued to the protruding taper. The nock must be installed so that the slot is at a 90-degree angle to the odd colored feather on

CONSTRUCTION OF FIBERGLASS AND ALUMINUM ARROWS

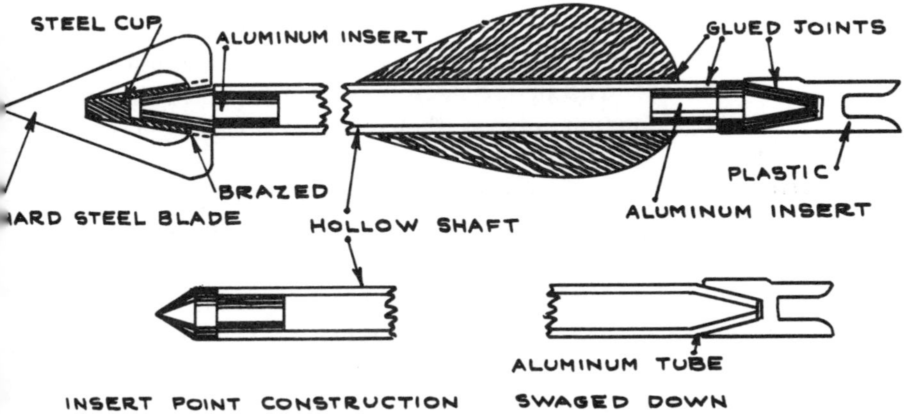

FIGURE 3

the arrow. This "cock feather" always faces the archer as the bow is drawn.

To those familiar with modern manufacturing techniques, it appears obvious that the cost of the arrow could be reduced by combining the front assembly and rear assembly into one component each. Actually this is sometimes done as shown in the two alternate construction sketches in *Figure 3*. The point is turned from a piece of solid steel on a screw machine. A conical tip is machined in the steel, and a cylindrical section cut on the rear to fit the inside of the hollow shaft. The inside of the steel is sometimes drilled out to reduce weight, or grooves are cut in the outside for the same purpose. The steel point is then force fitted inside the aluminum shaft, making a permanent assembly, or glued into a fiberglass shaft.

The rear end of an aluminum shaft may be fitted with an insert as shown in the upper illustration, or the rear end of the aluminum tube may be swaged down to form a conical surface. The plastic nock may then be glued directly to the aluminum tube as shown in the lower right-hand sketch.

Fiberglass arrows cannot be swaged down at the rear, so that a metal insert is used to provide a conical surface. It seems like a simple idea to mold the nock so that it has a cylindrical surface on the front which could be glued directly to the inside of the fiberglass tube. This is not a good construction, since it is not unusual for one arrow to strike another arrow already in the target, breaking the nock off. With a tapered insert it is a simple matter to cut off the damaged nock and glue on a new one. If the nock extended inside the shaft of the arrow it would be necessary to set up the damaged arrow in a lathe and machine out the damaged nock very carefully, or the arrow would be ruined.

Fiberglass arrows are generally recognized as the most rugged arrows available today. The shafts are straight, tough, and uniform. The accuracy is generally better than with a matched set of wood arrows, but not quite as good as the accuracy of aluminum arrows. If a fiberglass arrow is in the target and is struck by a following arrow on the side, a mark will be left, but virtually no other damage will occur. If a wooden shaft suffers the same treatment, a section of wood will be gouged out of the shaft, or the shaft will actually be broken.

Aluminum arrows provide the ultimate in accuracy. They are not recommended for rough field use, but they are superb arrows for target shooting. The normal course of development of an archer would be to start with wooden arrows, since the first set always takes a great deal of beating, and some arrows are lost due to inaccurate shooting. As shooting form improves and losses due to wild shots are eliminated, a fiberglass set of arrows provides an excellent means by which shooting techniques can be honed to a high degree of skill. If the shooter becomes serious about the ultimate in accurate shooting, aluminum arrows will provide the finest accuracy attainable. In shooting heavy game and small game, accuracy is of paramount importance; and the number of arrows fired is relatively small. Many top-notch archers use aluminum arrows for these purposes. Aluminum arrows do bend if struck a glancing blow; and, therefore, most hunters still use either wood or fiberglass arrows in field shooting where damage and loss are frequent. Even in target shooting, aluminum arrows will occasionally become bent. It is a good idea to make a small jig in which arrows can be rotated and checked for straightness.

ACCURACY OF ARROWS

Suppose that you are an average archer, and that you would like to compare the performance of several different types of arrows. If you measure the size of groups, you can receive very deceptive results. Supposing the set of arrows was generally very good, but had one arrow that flew to the left and a second that flew to the right. Any group measurements that included these two arrows, or either one, would give a totally erroneous picture of the performance of the over-all set. The same problem of measuring the quality performance of products is a very serious one in modern industry. Methods for mathematically evaluating the performance of things have been highly developed, and

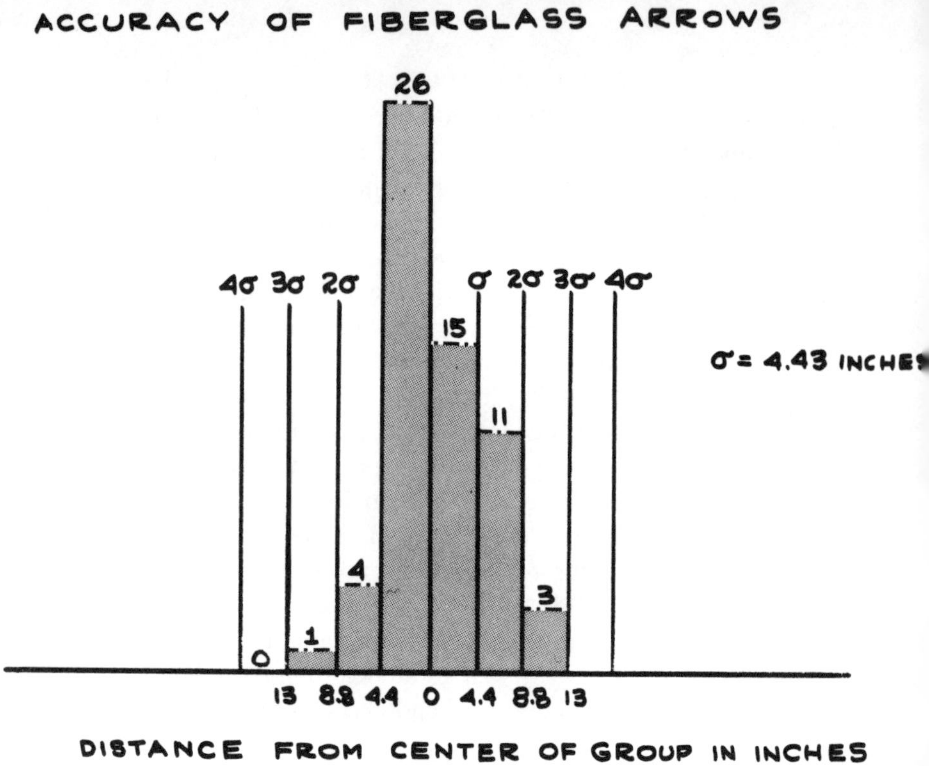

FIGURE 4

the application of these techniques to archery provides some very interesting results. The statistical techniques applied to a set of a dozen fiberglass arrows is shown in *Figure 4*. Five targets were fired of 12 arrows each, a total of 60 shots, at a range of 30 yards. The horizontal distance of each arrow from the aiming point was measured and recorded. The data was then analyzed to calculate a "standard deviation." One standard deviation is referred to as a "sigma." According to statistical theory, 67 per cent of the shots fired at any target should fall within the limits of plus or minus one sigma. In actual fact, 26 arrows fell within the area to the left of the center of the group within a distance of 4.43 inches, which is the one-sigma limit for these arrows. Fifteen arrows fell into the same area on the right side of the target. Thus, the total within a plus or minus one-sigma limit was 41 arrows, or 68 per cent of the total.

The procedure to find the "sigma" for a set of arrows is as follows: Draw a vertical line through the center of the bullseye of the target, and extend it as far as possible up and down. Shoot all the arrows in the set at the target—say 10—and measure the distance from *each* arrow to the *center line*. Record arrows to the right as plus, arrows to the left as minus. Then add up the column of figures; suppose that it comes to plus 30. This means that the *center* of the group is three inches to the right of the line. Now go back to your data and subtract three inches from each reading. These are your "residuals," or the distance from the impact of each arrow to the *center* of the *group*.

Repeat this procedure until your patience wears out. The larger the number of shots, the more accurate the results, but be sure to use the full set each time.

Standard Deviation (Sigma)

This standard deviation is calculated from the *residuals*. The formula is:

$$\text{Sigma} = \frac{\sqrt{\sum_{i=1}^{n} x_i^2}}{\sqrt{n-1}}$$

n = total number of arrows shot at target.
x_i = individual residuals starting with i = 1 for first arrow to i = n for last arrow.
Σ = summation.

This formula is in the mathematical language which may be simplified as follows:

Sigma is calculated by calculating the "residual" for

each arrow fired at the target. Each of the residuals is squared, and then the squares are all added up. You then take the square root of the sum.

This number is then divided by the square root of the number of arrows fired minus one. Thus, if a total of 50 arrows has been fired, then $\sqrt{50\text{-}1} = 7$.

A simplified example is shown below for *ten* arrows.

Shot No.	*Impact Point*	*Residual (x)*	*Residual Squared (x^2)*
1	+6	+3	9
2	+2	—1	1
3	+5	+2	4
4	—3	—6	36
5	+6	+3	9
6	—1	—4	16
7	+7	+4	16
8	—2	—5	25
9	+9	+6	36
10	+1	—2	4
	Sum = +30		Sum = 156

Since the sum of the impact points is plus 30, it means that the center of the group is three inches to the right of the center of the target.

The residuals are calculated by subtracting three from each impact point. The fourth column shows the squares of the residuals, which are all positive. The sum of the squares of the residuals comes to 156. The sigma is calculated by:

$$\text{sigma} = \frac{\sqrt{\text{sum of squares of residuals}}}{\sqrt{\text{number of shots fired minus one}}}$$

$$\text{sigma} = \frac{\sqrt{156}}{\sqrt{10-1}} = \frac{12.5}{3} = 4.16''$$

To find the number of arrows falling within the sigma limits, be sure to use the residuals, which are the distance from the center of the group.

NORMAL DISTRIBUTION CURVE

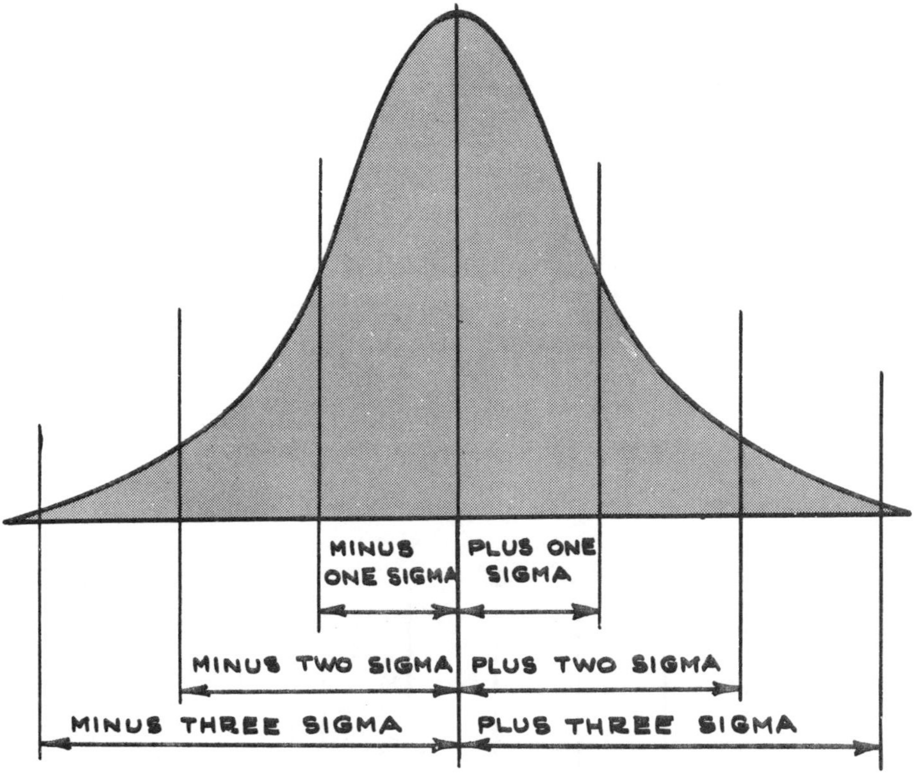

	SIGMA IN INCHES	PERCENTAGE OF ARROWS WITHIN ± ONE SIGMA	WITHIN ± TWO SIGMA	WITHIN ± THREE SIGMA
THEORY:		67	95	99.8
ACTUAL:				
FIBERGLASS SET	4.4	68.4	93.4	100
GOOD CEDAR SET	7.4	60	96	100
POOR CEDAR SET	12.8	70	93.5	100

FIGURE 5

In this case, seven arrows or 70 per cent fall within the plus or minus one-sigma limit. All of the arrows fall within the two-sigma limits.

A normal distribution curve with the results of these tests is shown in *Figure 5*. This Figure compares the theoretical values given by statistical theory with the actual results found in the simple examples given in this chapter.

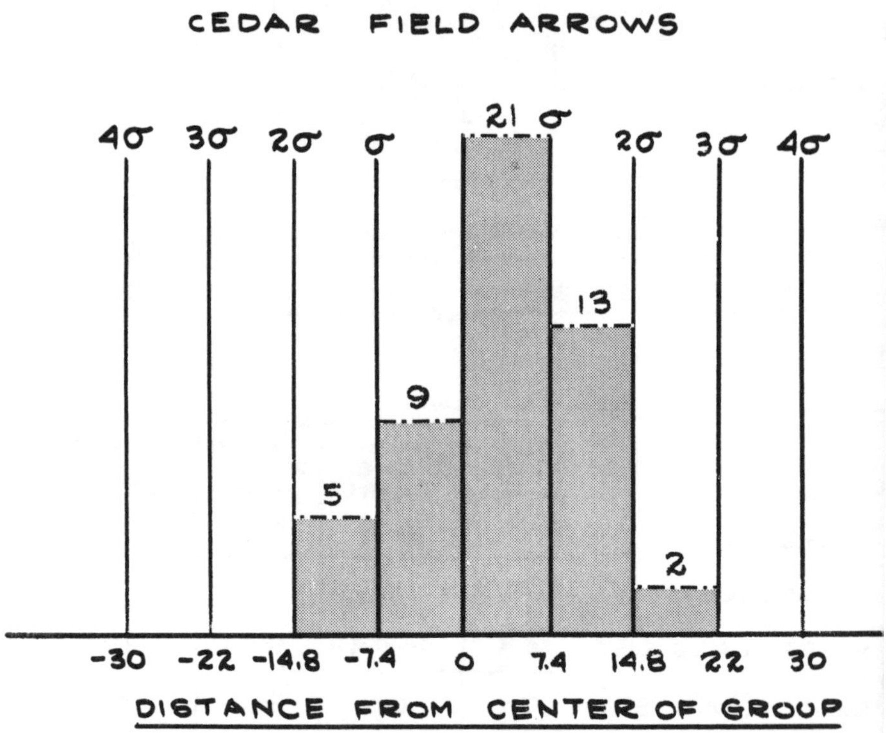

FIGURE 6

Tests with Wood Field Arrows

The actual results of tests with the same shooter at 30-yards range with a matched set of cedar field arrows is shown in *Figure 6*. In this case, the sigma has increased to 7.4 inches. Since the shooter error is the same amount in each of these tests, the standard deviation becomes an accurate means by which to compare the performance of the arrows.

Actual tests with fiberglass and aluminum arrows with mechanized systems to launch the arrows very accurately have given extremely small groups. Machine-fired groups with top quality arrows have given these results:

Target Arrows	*30 Yards*	*70 Yards*
Fiberglass Arrows	2⅛ inches	2⅝ inches
Aluminum Arrows	1⅛ inches	5 inches

Inaccurate Arrows

Tests with an inaccurate set of arrows are shown in *Figure 7*. In this set there were two arrows which consistently flew to the right, and one that flew a long way to the left. The result is that the entire distribution curve becomes distorted. The sigma for this set of a dozen field arrows was 12.8 inches, a clear measure of very poor accuracy. Consider a shooter using these arrows at a target, and trying to learn to improve his techniques. One shot flies to the right, so he corrects to the left; the next arrow happens to deviate to the left, and his shooting becomes wilder as he tries to correct successively for the inaccurate performance of the arrows themselves.

Once you have established that a set of arrows is inaccurate, that set should be immediately discarded for any

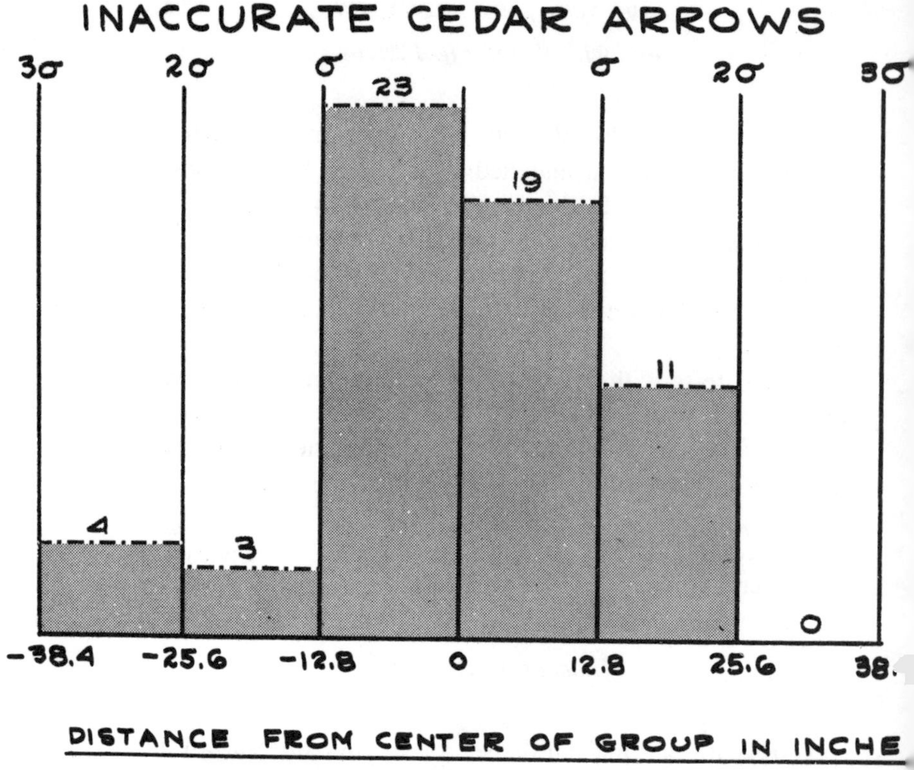

FIGURE 7

serious shooting. If the arrows have been numbered, you can analyze their performance on half a dozen targets and jot down which ones fly to the left, which ones fly to the right, and which ones are accurate. If you carefully segregate them into these three classes, they may be used for field shooting with fair effectiveness, but no attempt should be made to do any serious target shooting with such arrows.

Numbering Arrows

If you are seriously interested in improving your archery accuracy, then number your arrows. The best location is right in front of the cock feather, where it is readily visible as you place the arrow on the string. The advantage of numbering can be seen by referring to the simplified example used for calculating the "sigma." Ten arrows are listed. As the position in the target is noted, you see that arrows No. 7 and No. 9 fly to the right, and No. 4 and No. 8 are farthest to the left. These may be bad shots, or it may be due to the arrows. If you note the arrows farthest to the left and right on a half-dozen targets, you can estimate very accurately the position of each arrow on the target, if you make no aiming errors.

A plot of the average point of impact for each arrow in the inaccurate set of cedar field arrows is shown in *Figure 8*. This data may be calculated right from the same

AVERAGE IMPACT POINTS OF INACCURATE ARROWS

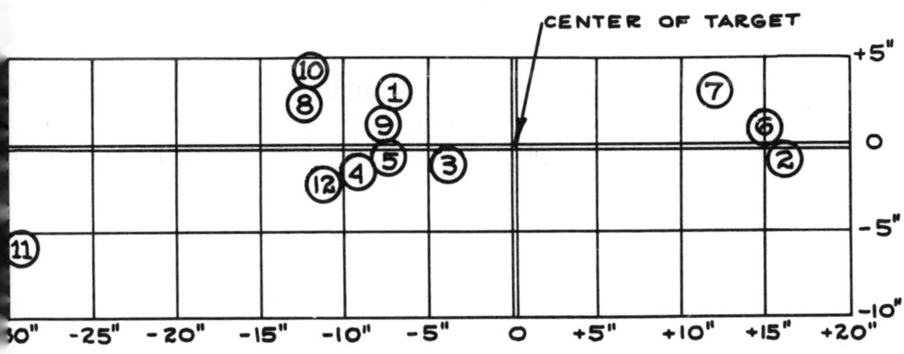

FIGURE 8

sheets used to calculate the "sigma." Add up the impact points for all the shots with arrow No. 1, and divide by the number of shots. Repeat this with each successive arrow, and plot their position. By using such a chart, and compensating for each arrow in your point of aim, you can *cut your group size in half*. The set of arrows shown in *Figure 8* has three arrows that hit far to the right, and one that flew far to the left. In order to get any reasonable accuracy out of this set, arrow No. 11 was discarded. The bow sight was set to properly center the large number of arrows on the bullseye of the target. When arrows 2, 6, or 7 were used, the bow was aimed out in the black (3) ring way to the left. This kind of shooting will greatly improve the performance of an inaccurate set of arrows, but a better solution is to buy a set of accurate arrows in the first place. The accuracy of your arrows should be checked periodically, as both wood and aluminum arrows get bent, and must be straightened to restore their accuracy.

ARCHERY AIMING SYSTEMS

There are three systems for aiming a bow and arrow in wide use today. Most archers start off with the "instinctive" aiming approach. The archer focuses on the target to be hit, and lets the arrow fly in that general direction. With practice, accuracy improves. This type of aiming is widely used in hunting where every shot is at a different range and angle. Instinctive aiming is the technique used in throwing a baseball, and it really is using the computer in your brain to balance all the factors involved and make the minute muscular adjustments needed to hit the target. Instinctive aiming requires a great deal of practice to develop proficiency, but exhibition archers using this technique can perform amazing feats. Instinctive aiming is recommended for the hunter who is willing to spend a lot of time developing and maintaining proficiency.

Point of Aim

The second technique utilizes a marker on the ground between the archer and the target. The archer lines up the tip of his arrow with the aiming point, and lets fly. If the arrow strikes too high, then the aiming point is moved closer to the archer, which has the effect of lowering the angle of the bow. The point of aim technique has been widely used in target archery for centuries, and it is a common technique in archery classes at schools and colleges.

This technique is simple to use, and can be applied to any bow and arrow. It has the disadvantage of requiring trial and error shifting of the aiming point, and, of course, cannot be applied to field shooting or hunting. It is an excellent intermediate step in learning archery between the instinctive aiming technique and the use of sighting equipment.

BOW SIGHTS

For maximum accuracy, most modern archers use some form of bow sight. This may range from a long pin stuck crosswise through a piece of tape on the front of the bow up to a full tournament sight with calibrated vertical scales on the front and back of the bow and two reticules.

The reason for the wide range of sights may be seen by considering a rifle sighting system. The barrel of the gun establishes the direction of the bullet's flight very exactly. With good equipment you can pump bullets into a two-inch diameter circle at 100 yards all day long. In order to properly aim the barrel, a rigid sight is attached at the front, and a movable sight attached at the rear. Slight movements of the rear sight are all that are required to have the bullets strike the bullseye at ranges of 25 yards to 300 yards. Normally, a quarter-inch change in rear-sight elevation will more than cover this change in range.

The problems in accurately aiming a bow and arrow are more complex. There is no tube to accurately launch the arrow on each shot. Instead, the front of the arrow rests vertically on an arrow rest attached to the bow, and is positioned horizontally against the striker plate in the sight window. As the arrow is released, the string pushes the arrow against the bow horizontally, and the arrow actually curves around the bow on its way to the target. If the arrow is a proper stiffness for the bow, it will bend uniformly on each shot and give good accuracy.

Another complexity with the archery system is that the nock of the arrow can be placed anywhere within a vertical range of about two inches on the string, and this will introduce wide variations in the launch angle of the arrow. This problem is solved by wrapping a fine white or contrasting thread on the bowstring to provide a place to nock the arrow on every shot. The nocking point should be located so that the arrow is at right angles to the string with the bow at rest, or within one quarter of an inch *above* this position. Be sure it is not low.

Drawing the Bow

In drawing the bow, it is important that the same *length* of draw be used on each shot. A short draw means

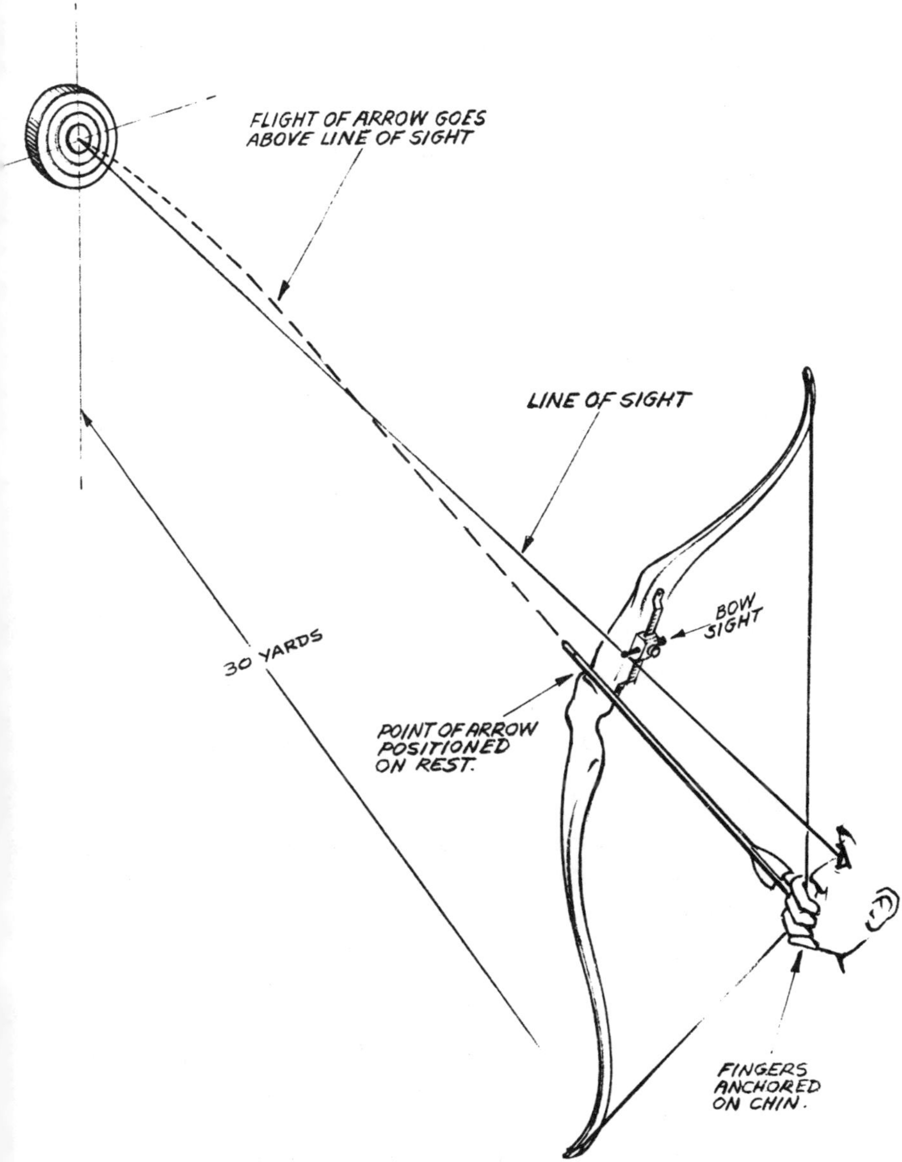

FIGURE 9

IRVING ALLEN

a low velocity and a *low* shot. A longer-than-usual draw will give a high velocity and a high shot.

If you draw so that the bow string is in firm contact with the tip of your nose on each shot, then you have a good guide on the length of pull. Centering the string on your nose also gives a good lateral positioning of the bow string.

The sighting geometry using a simple bow sight is shown in *Figure 9*. The bow sight will be described in more detail later on, but it basically consists of a small brass knob that can be positioned vertically and laterally on the bow. When the sight is properly adjusted, you lift the fully drawn bow until the brass ball is centered over the bullseye of the target, and release your shot. The arrow then rises above your line of sight and gracefully arches right into the bullseye of the target.

In order to achieve high accuracy in archery, the arrow must be in exactly the same place on the bow, and the bow must be exactly aligned with the target shot after shot.

This goal is accomplished with a bow sight as follows:

Elevation:

The front of the arrow is positioned on the arrow rest. The nock is positioned on the bow string by the "nocking point" in contrasting color fastened to the string. The arrow is gripped by the fingers, which in turn are brought to the same point on the chin on each shot. The last element is the vertical positioning of the bow sight on the bow. The elevator is adjusted so that when the eye superimposes the sight aiming point over the bullseye, the elevation is correct for the range.

To raise the point of impact, the sight point is *lowered* on the bow.

Windage:

Windage refers to the horizontal positioning of the arrows on the target. The front of the arrow is against the sight window on the bow. The rear of the arrow is positioned by the nock on the string. The string is held by the archer so that it is right in the center of the nose. The final adjustment is made in the horizontal screw of the bow sight, which swings the bow horizontally in relation to the target to have the arrows strike the center of the target. As a general rule, stiffer arrows fly farther to the left, and require a greater windage adjustment. Aluminum and fiberglass arrows of the same stiffness rating (known as "spine") need different windage adjustments.

Telescopic Sights:

A popular sighting system for rifles is the telescopic sight, with crosshairs which are aligned with the target. The telescopic sight reduces the error of aim to a very minute value, less than a *quarter of an inch at 100 yards.* This same accuracy of sighting can be applied to a bow. New telescopic sights designed specifically for bows have recently been announced on the market. The telescopic sight will align the bow with the target very exactly, both horizontally and vertically.

The arrow must still be positioned in exactly the same relationship to the bow on each shot to achieve accuracy. The actual sighting situation is the same as that shown in *Figure 9,* with a telescopic mounted on the bow in place of the bow sight. In effect, the bow sight "double checks" the uniformity of the other factors, and helps in further refining the accuracy of the system.

Tournament Sights:

Many tournament archers use complicated mechanical sights which achieve much of the accuracy of the telescopic sight. These consist of two vertical bars, one spaced in front of the bow, and one behind the bow. One adjustable housing is fitted on each bar, and a horizontal screw is fitted to each housing. When this is all adjusted, the archer has two aiming points which provide an aiming axis just like a telescopic sight. Very often a "kisser button" is fastened to the bow string a short distance above the "nocking point." The kisser button is touched to the lips on each shot to give a very accurate vertical positioning of the arrow.

Using a Bow Sight

Archery with a bow sight takes some practice. At first you will do worse with the bow sight than without it. A few days later, you will do equally well with and without, and from then on the bow sight will be a constant aid to better accuracy. One of the greatest benefits of the bow sight is that it helps you to see what you are doing wrong. If you flipped the bow to one side or the other, or released at the wrong time, you can predict where the arrow will go, and learn to do better next shot. If you released properly, and the arrow still flies to one side or the other, then check to see if your hold is proper. As a last resort, check to see if the arrow is bent.

A simple but accurate bow sight is shown in *Figure 10.* This sight is superior to most of the commercial sights on the market in two ways. A steel scale is fastened to the vertical frame, and this allows exact changes in elevation. Unlike rifle sights, a bow sight must be changed with each change in range, even if it is only ten yards. A slight change in setting is usually needed if a wind is blowing. Thus, you may find that aluminum arrows at 30 yards require a setting

FIGURE 10

of 2½ on the vertical scale, while fiberglass arrows require 1¼ inches at 50 yards. With these settings written down, you can get on the target with the first arrows shot.

It is also easy to figure out how far your group is shifted by a change of ⅛ inch on the sight. In shooting at 50 yards, the shift is about 7½ inches for each ⅛ inch shift on the sight. At 30 yards, the shift is about 4½ inches, although these figures will vary slightly, depending on your length of draw.

The assembly of the sight is shown in *Figure 11*. The sight is a small ball at the end of a horizontal screw. This is threaded into a "Block" which is cross pinned to the "Housing." The assembly is moved up or down the "Frame" to select the right elevation. When a position has been selected, it is locked in place by the "Lock Screw" which bears on a small sheet metal "Shoe." The "Shoe" can be eliminated, but if this is done, the "Lock Screw" will tend to leave circular marks on the "Scale." The "Shoe" is a simple thing to make, and a useful addition.

The windage is changed by threading the "Horizontal Screw" in the "Block." One way to measure windage is the number of exposed threads between the "Housing" and the head of the screw. The threaded construction allows very fine adjustment of windage, and a filed notch on the head of the screw can serve as an easy index for fine adjustments.

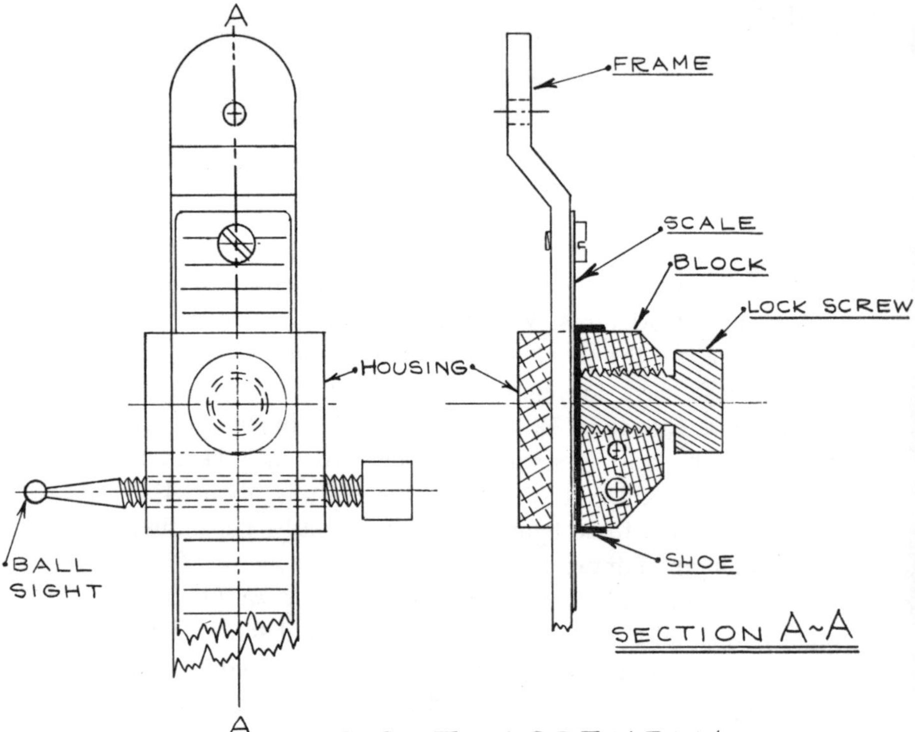

SIGHT ASSEMBLY
FIGURE 11

The parts making up the sight are shown in *Figure 12* through *Figure 16*. Almost any materials available can be used, including brass, aluminum or steel. The frame should be steel, or a hard-tempered brass, so that it is not easily bent.

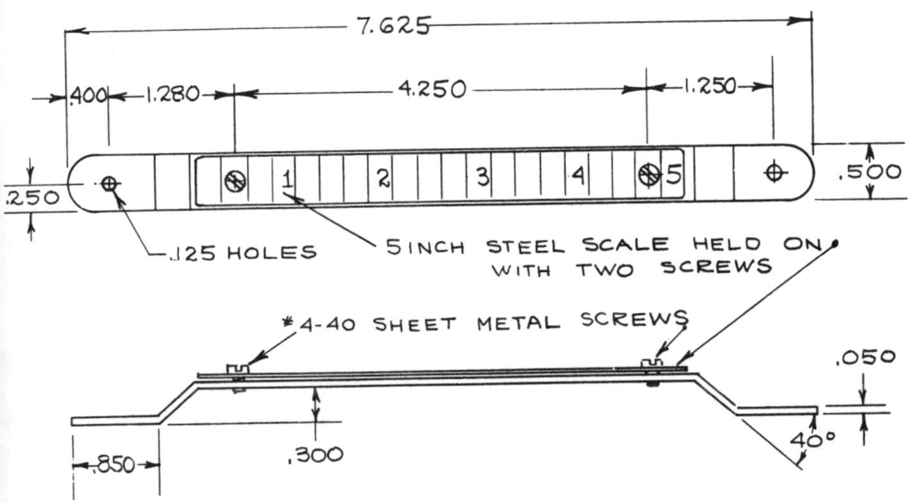

MATERIAL: STEEL

FRAME

FIGURE 12

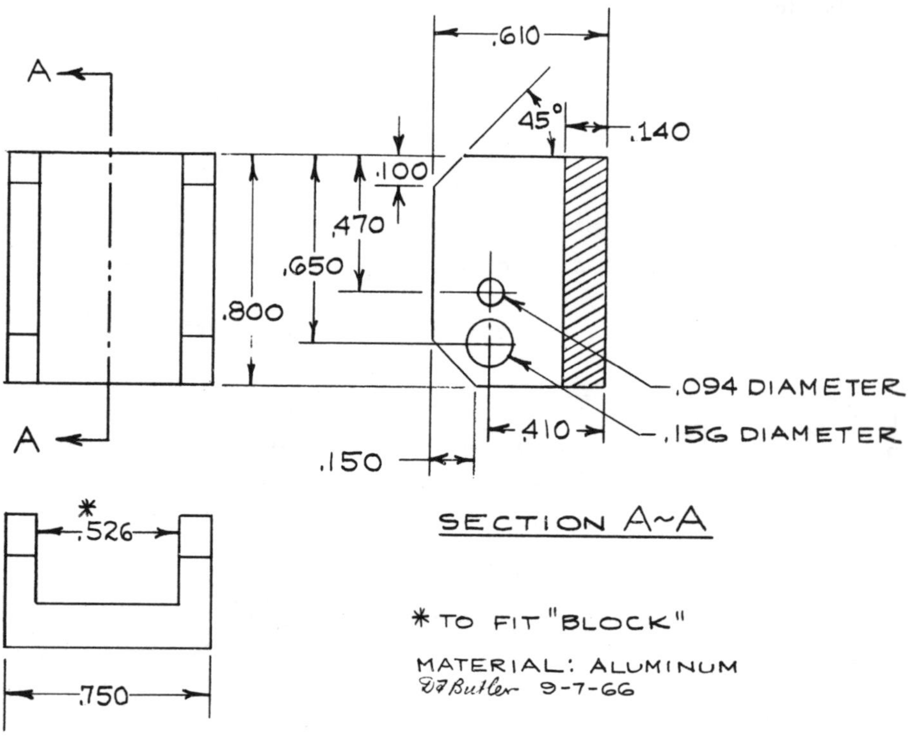

HOUSING

FIGURE 13

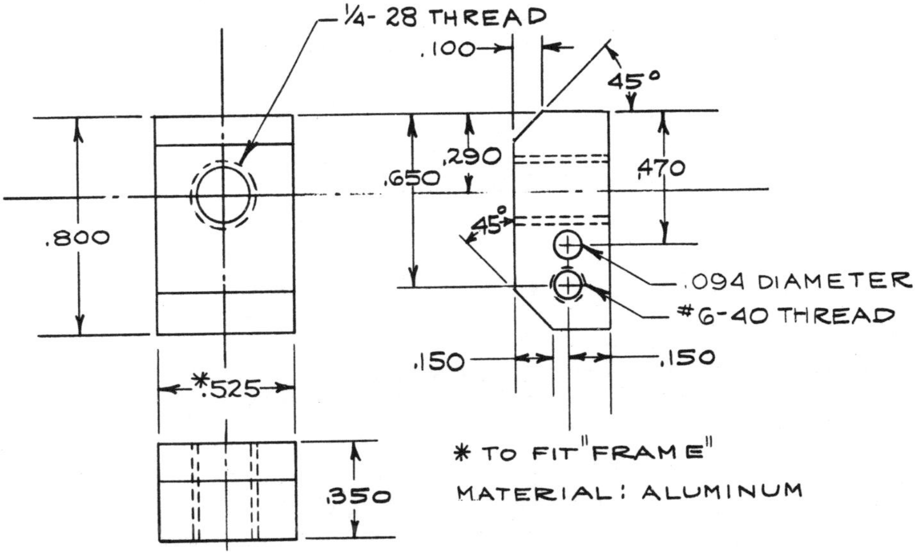

BLOCK

FIGURE 14

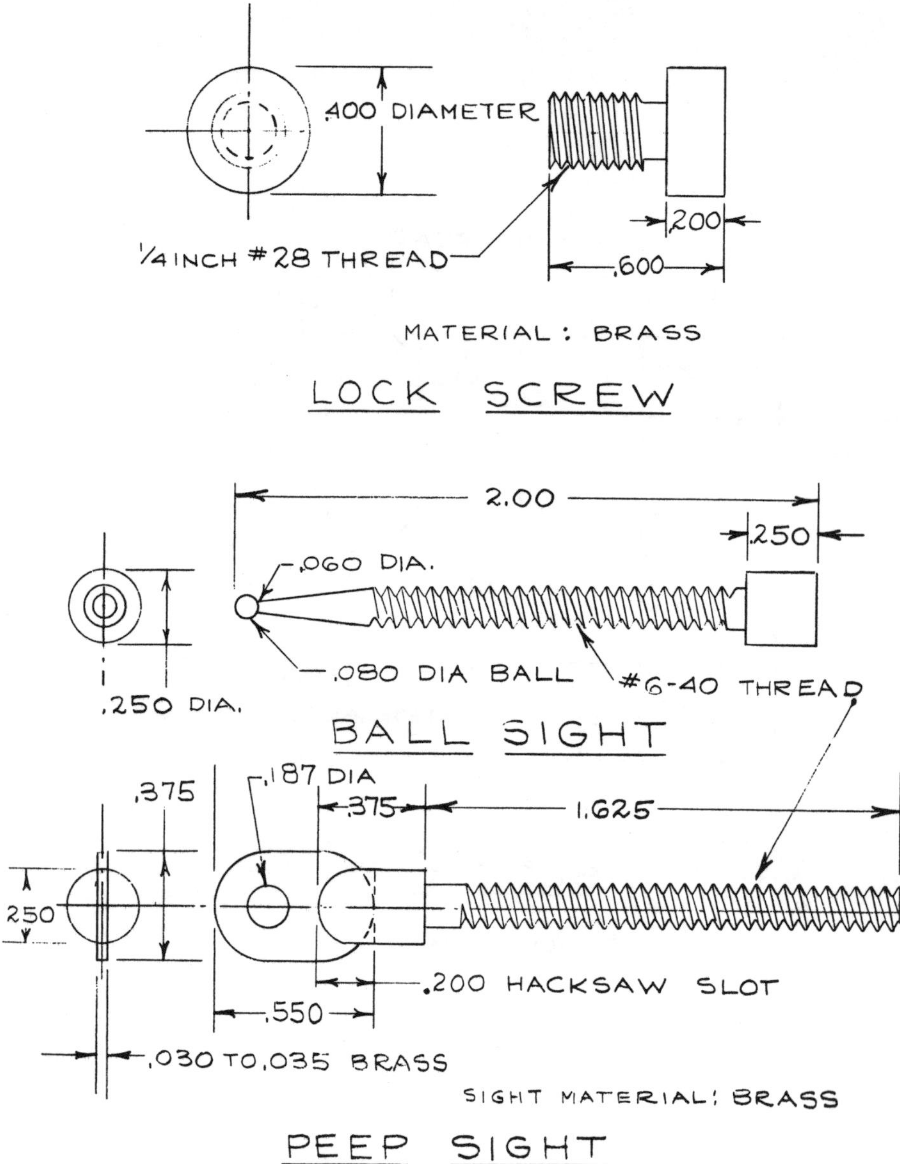
.400 DIAMETER
¼ INCH #28 THREAD
.200
.600
MATERIAL: BRASS
LOCK SCREW
2.00
.250
.060 DIA.
.080 DIA BALL
#6-40 THREAD
.250 DIA.
BALL SIGHT
.375
.187 DIA
.375
1.625
250
.200 HACKSAW SLOT
.550
.030 TO .035 BRASS
SIGHT MATERIAL: BRASS
PEEP SIGHT

FIGURE 15

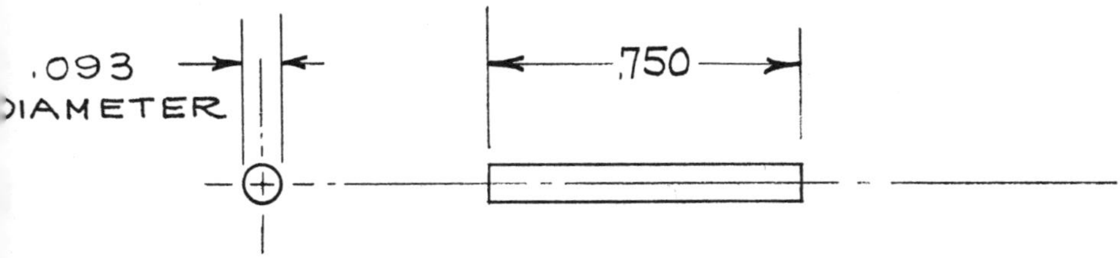

MATERIAL: STEEL

CROSS PIN

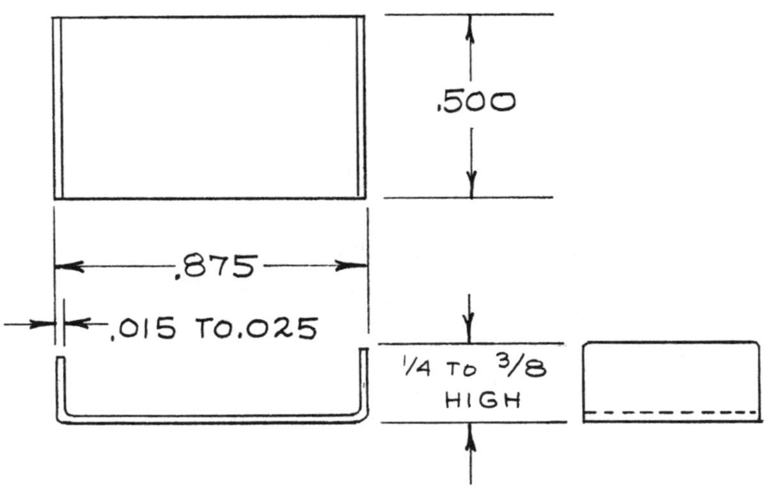

MATERIAL: COPPER

SHOE

FIGURE 16

For the sight shown in *Figure 11,* the following materials were used:

Frame: Steel, covered with polyethylene in center section.
Scale: Stainless steel 6 inch scale cut down to 5⅛ inches.
Screws: Holding "Scale" to "Frame": Self Tapping #4-40 screws.
Housing: Aluminum. This is easy to machine and can be made by hand if care and patience are used. The inside width of the housing should be only about .010″ to .015″ [1/64″] wider than the "Frame."
Block: The "Block" for these sights was made from aluminum. Brass or steel may also be used, but aluminum is strong enough, is easy to work, and does not rust. The "Block" should be a nice slide fit inside the walls of the "Housing", so go slowly as you machine or file the final dimensions.
Cross Pin: Any wire rod of the proper diameter will do. Drill rod is excellent, if available.
Lock Screw: Brass is an excellent material, for it is easy to work, and gives a very classic contrast to the aluminum "Block" and "Housing."
Shoe: Copper was chosen, since it is very easy to work. The "Shoe" may be formed up around the "Block" in a vise.
Ball Sight: The "Ball Sight" should be turned up from a one-quarter-inch diameter rod of brass. It can also be made from a two-inch long No. 6-40 brass machine screw, but this will have a very small head. A No. 6-40 brass nut can be run up the thread and tightly jammed under the head to give more surface to grip. The "Ball Sight" should always be chosen for the first archery sight, for it is the easiest type to learn to use.
Peep Sight: This sight is assembled from a 1¼-inch

brass rod, and a small plate. A short length of the rod should be left the full ¼-inch diameter, and a hacksaw slot should be cut into the top for a depth of .200 inches. The small sheet of brass should be soldered into the slot.

The best size for the peep sight hole will vary with the shooter and the range. This design was suggested by Mr. G. Randolph Erskine of Wallingford, Connecticut. As an archery coach he has found that best accuracy occurs when the diameter of the peep hole allows you to see the bullseye on the target, and a small portion of the seven ring. This gives your eye the best optical image to center the bullseye very closely. The author has found that a .156-inch diameter peep is about right for a range of 50 yards, and the .187-inch peep shown in *Figure 15* is a good size for 30 or 40 yards. It is important that the peep hole be located on the axis of the threads, or there will be a slight change in elevation as the sight is adjusted for windage.

Assembly Of The Bow Sight

The sight is assembled in the following sequence: Cut off the six-inch steel ruler to slightly over five inches, and fasten it to the "Frame" with the two small self-tapping screws. The holes in the ruler should be the full diameter of the threads, and the holes in the "Frame" smaller to give the threads a good bite.

Slip the "Housing" underneath the "Frame." Place the "Shoe" around the "Block," and slide the assembly in between the walls of the "Housing." The assembly is held together by the small "Cross Pin." The "Lock Screw" should be threaded in place, and tightened lightly. The bow sight should be mounted on the bow with two small screws before threading in the "Ball Sight."

Two installations of these bow sights are shown in *Figure 17*. The bow on the left is a very light target bow, rated at 23 pounds at a 28-inch draw. The sight is installed on the back of the bow (the side towards the target) with two small wood screws. Small holes should be drilled

Installation of bow sights on two bows. The sight is mounted on the side facing the target on the black bow, and on the side facing the archer on the near bow.

FIGURE 17

through the fiberglass and into the riser section of the bow before driving the screws.

The bow on the right side of the illustration is a heavier bow, requiring 36 pounds at a 28-inch draw. This bow has a complex shape on the back, but a smooth side facing the archer, so the sight has been installed on the face. The sight was originally held on this bow with tape, but the sight tends to shift, particularly in hot weather. A small pad of electrician's tape at the points of contact between the bow and the sight help to make a firm joint even with relatively small screws.

After the sight is fastened to the bow the "Ball Sight" should be threaded in. This sight does not require a lock nut to hold an adjustment, but a nut can be threaded over the ball and locked against the housing if this added security is desired. Another way is to wrap some adhesive tape around the threads under the head of the sight to cause some friction as the sight is moved laterally.

The Bow Sight in Use

A young archer is shown using the bow sight on a lightweight bow in *Figure 18*. Use of a light bow allows the young man to fully draw the bow to a proper anchor position for each shot. He can hold steadily enough to use the bow sight properly. The ball sight is the best design to start with, and should be used for several months before shifting to a more complex sight. Many archers find the ball sight an excellent all-around sight for field shooting and informal target shooting.

With a ball sight the entire target is in view at all times. No one can hold a bow absolutely steady, and with a ball sight you can see the sight drifting towards the bullseye, and release as you are coming on center. If the bow is

Young archer using a bow sight.

FIGURE 18

moving significantly when you release, the arrow will generally miss the bullseye.

The peep sight offers the next level of aiming accuracy, and a new level of frustration. The motion of the sight drifting back and forth across the target is much more apparent. If the sight is properly aligned the bullseye is nicely centered, but when you have drifted you see a half bullseye, which is obviously wrong. The peep sight will differentiate more clearly between a good hold, and a bad one. This discrimination can be a real help in coaching yourself not to release arrows when you are "almost on" the target.

There are a number of more complex commercial sights for the advanced archer. Many of the commercial reticules, such as a crosshair, or post with bead, can be adapted to your homemade sight. Very few commercial sights give a reference scale for elevation, and this is important. Commercial sights are available with many additional features, such as small bubble levels which warn you if you are tipping the bow. Examine a sight before purchase, and assure yourself that it is rugged enough to hold adjustments during normal archery use.

CONCLUSION

The final result of using good archery equipment is shown in *Figure 19*. An average weekend archer can develop his or her skill to the point where he can keep all of the arrows on a three-foot diameter target at *50* yards with only one hour a week of practice. The secret is not long and arduous muscle-building marathons, but rather the choice of proper equipment, and care and intelligence in its use. In choosing a bow, be sure to compromise in the direction of a *lighter* bow. Start with wooden arrows from open racks to develop your form, but switch to a set of tough, accurate,

The end of a successful archery session.

FIGURE 19

fiberglass arrows when you can raise the cash. Start with simple equipment, and add better accessories as you master the simple ones.

One way to start is to use a bare bow and wood arrows at a three- or four-foot diameter target at 30 yards until all of the arrows stay on the target, and the scores average fifty for ten shots. Then increase the range or add a new piece of equipment to raise your level of skill. Accept the fact that new equipment will cause a drop in scores temporarily, but persist until you master the new techniques. Two of the most frustrating new pieces of equipment are the three-fingered archery glove, and the bow sight. The stiff leather on the three-fingered glove will cause dropped arrows for several days. The problem is most acute with aluminum target arrows, which are very small in diameter. The bow sight will cause a drop in scores for a week. During the second week scores will equal bare bow scores, and from then on the bow sight scores will climb steadily.

If the archery range can be set up and left in place progress will be rapid. Thirty arrows in an evening after work is only fifteen minutes practice, yet this will improve techniques at a rapid rate. One of the relaxing things about this sport is that you can advance at your own speed, and in your own direction. If your interest is in hunting, then animal target faces can be placed over the target for more realistic practice. If your interest runs to indoor tournaments, then smaller and smaller targets may be used to improve skills at 20 or 30 yards. If you like to watch arrows arc gracefully through the air, then use a three- or four-foot target, and gradually increase the range to 50 yards or beyond.

Index